HARDPRESS.NET
HOME OF HARD-TO-FIND BOOKS

Elements of Physics
by Neil Arnott

Address:
HardPress
8345 NW 66TH ST #2561
MIAMI FL 33166-2626
USA
Email: info@hardpress.net

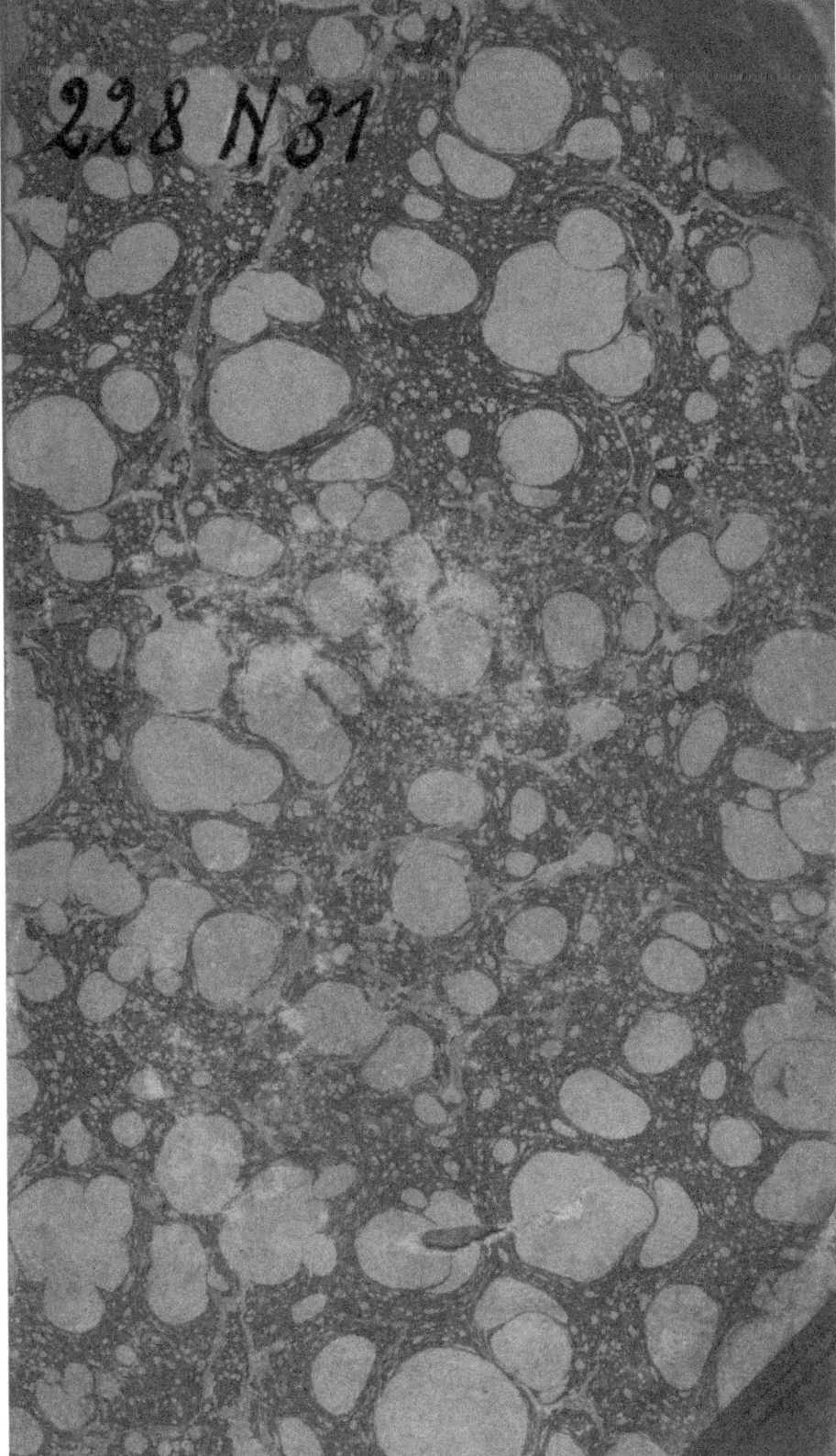

PRESENTED
by the
CLOTHWORKERS COMPANY
OF
London,
To *James Elliott*
at the annual examination of the
SCHOLARS
in their
Free Grammar School
OF
Sutton Valence, Kent,
for his proficiency in
Mathematics
on the *29 June 1849*
Geo. Brausfield Master of the Company.

Rev. H. M. MILLIGAN, B.A. *Head Master of the School*

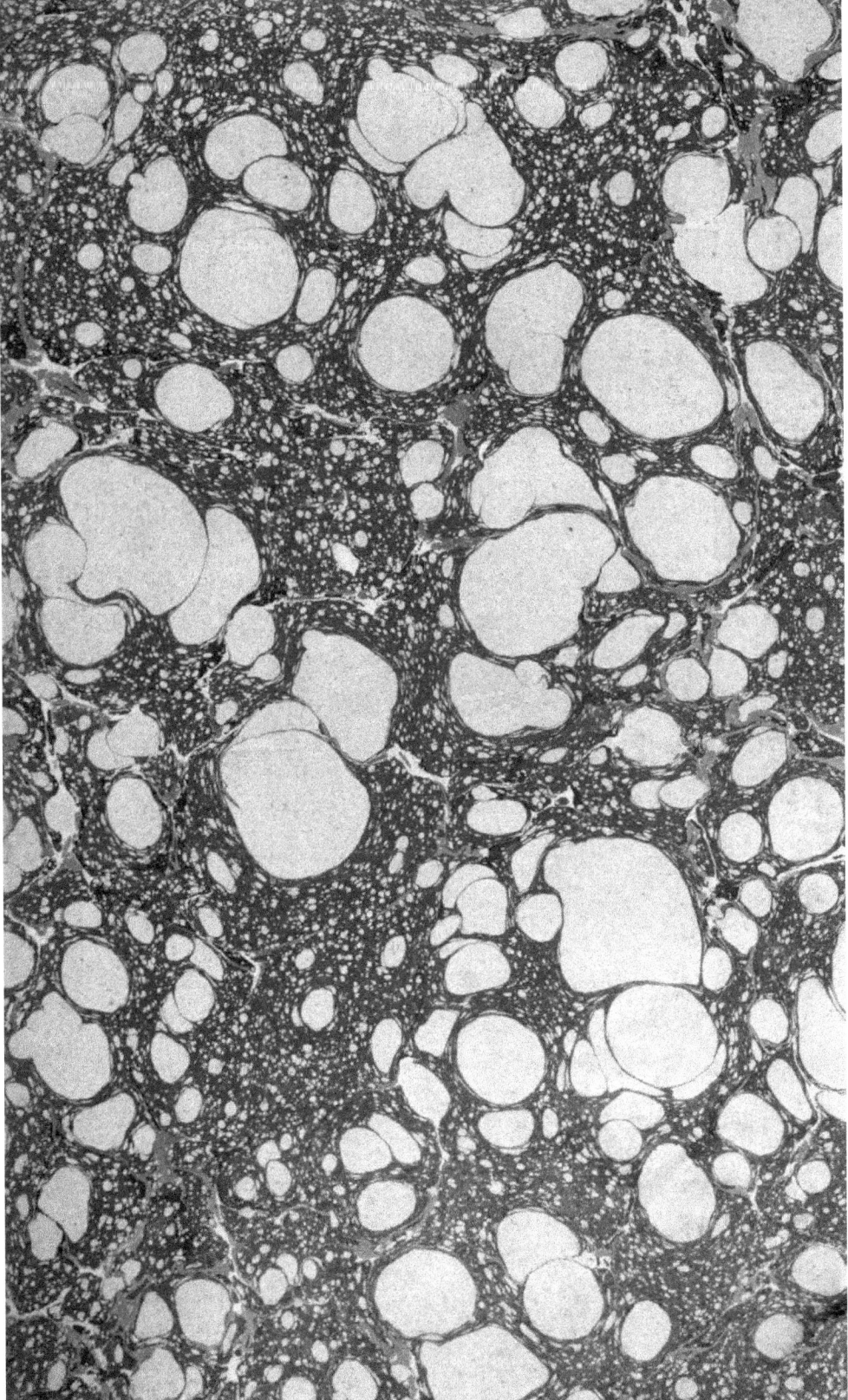

228 N 31

ELEMENTS

OF

PHYSICS,

OR

NATURAL PHILOSOPHY,

GENERAL AND MEDICAL,

EXPLAINED INDEPENDENTLY OF

TECHNICAL MATHEMATICS.

IN TWO VOLUMES.
VOL. II.—PART I.

COMPREHENDING THE SUBJECTS OF

HEAT AND LIGHT.

By NEIL ARNOTT, M.D.,

OF THE ROYAL COLLEGE OF PHYSICIANS.

LONDON:

PRINTED FOR LONGMAN, REES, ORME, BROWN, AND GREEN,
PATERNOSTER ROW; AND
T. AND G. UNDERWOOD, FLEET STREET.

1829.

LONDON
PRINTED BY J. L. COX, GREAT QUEEN STREET,
Lincoln's-Inn-Fields.

ADVERTISEMENT

THE second part of volume ii, comprising the subjects of ELECTRICITY, MAGNETISM, and ASTRONOMY, and concluding the work, will be put to press after the publication of the present part. The first volume was originally published without the second, although the whole manuscript was prepared, because other works on Natural Philosophy were offered to public notice about the time. The delay of two years with respect to the second volume has occurred, because the author's very little leisure from the duties of his profession (than which perhaps none more interestingly absorbs the time and faculties) was completely taken up by attending to the repeated calls for editions of vol. i. A friend, however, has superintended the printing of the last edition, and has allowed him to proceed with vol. ii. These explanations are given as an apology to the many persons who have honoured the work by expressing disappointment at the tardy appearance of vol. ii.

The author, while preparing the fourth edition

of vol. i, received a copy of a French edition, in which the translator, M. Richard, to fit the work for the general use of public schools and colleges in France, had given in notes the common algebraical formulæ for the various cases described. The author, at one time, intended to have done this himself, but afterwards determined only to add a few remarks on the subject at the end of vol. ii. To this determination he still adheres. In one of the North-American English editions of the work, there are also copious notes, but as the author has not yet been able to procure a copy, he cannot remark upon them.

In the present or fourth edition of vol. i, the subject of speech is still further analyzed, by a complete explanation of the hitherto unknown nature of the defect called *stuttering* or *stammering* ; and the discovery of its nature has suggested to the author an effectual remedy, so simple, that sufferers in general will be able at once to adopt it from the description now given. That the purchasers of former editions may not be obliged to procure the last on this account alone, the chief additions to the section in vol. i. are here subjoined. They occur at page 610 of the fourth edition ; and they should be inserted at page 565 of the first edition, at page 589 of the second, and at page 598 of the third edition.

London, November 1829.

" The most common case of stuttering, how-ever, is not, as has been almost universally believed, where the individual has a difficulty in respect to some particular letter or articulation, by the dis-obedience, to the will or power of association, of the parts of the mouth which should form it, but where the spasmodic interruption occurs alto-gether behind or beyond the mouth, *viz.* in the glottis, so as to affect all the articulations equally. To a person ignorant of anatomy, and therefore knowing not what or where the glottis is, it may be sufficient explanation to say, that it is the slit or narrow opening at the top of the windpipe, by which the air passes to and from the lungs—being situated just behind the root of the tongue. It is that which is felt to close suddenly in hiccup, arresting the ingress of air, and that which closes, to prevent the egress of air from the chest of a person lifting a heavy weight or making any strain-ing exertion ; it is that also, by the repeated shutting of which, a person divides the sound in pronouncing several times, in distinct and rapid succession, any vowel, as o, o, o, o. Now the glottis during common speech need never be closed, and a stutterer is instantly cured if, by having his attention properly directed to it, he can keep it open. Had the edges or thin lips of the glottis been visible, like the external lips of the mouth, the nature of stuttering would not so long have remained a mystery, and the effort neces-sary to the cure would have forced itself upon the attention of the most careless observer ; but be-cause hidden, and professional men had not detected in how far they were concerned, and the patient himself had only a vague feeling of some difficulty, which, after straining, grimace, gesticulation, and sometimes almost general convulsion of the body, gave way, the uncertainty with respect to the subject has remained. Even many persons who by attention and much labour had overcome the defect in themselves, as Demosthenes did, have not been able to describe to others the nature of their efforts, so as to ensure imitation: and the author doubts much whether the quacks who have succeeded in relieving many cases, but in many also have failed, or have given only temporary relief, really understood what precise end in the action of the organs their imperfect directions were accomplishing.

" Now a stutterer, understanding of anatomy only what is stated above, will comprehend what he is to aim at, by being farther told, that when any sound is continuing, as when he is humming a single note or a tune, the glottis is necessarily open, and therefore, that when he chooses to begin pronouncing or droning any simple sound, as the *e* of the English word *berry* (to do which at once no stutterer has difficulty) he thereby opens the glottis, and renders the pronunciation of any other sound easy. If then, in speaking or reading, he joins his words together, as if each phrase formed but one long word, or nearly as a person joins them in singing (and this may be done without its being at all noted as a peculiarity of speech, for all persons do it more or less in their ordinary conversation), the voice never stops, the glottis never closes, and there is of course no stutter. The author has given this explanation or lesson, with an example, to a person, who before would have required half an hour to read a page, but who immediately afterwards read it almost as smoothly as was possible for any one to do ; and who then, on transferring the lesson to the speech, by continued practice and attention, obtained the same facility with respect to it. There are many persons not accounted peculiar in their speech, who in seeking words to express themselves, often rest long between them on the simple sound of *e* mentioned above, saying, for instance, hesitatingly, "e I e......think e...... you may,"— the sound never ceasing until the end of the phrase, however long the person may require to pronounce it. Now a stutterer, who to open his glottis at the beginning of a phrase, or to open it in the middle after any interruption, uses such a sound, would not even at first be more remarkable than a drawling speaker, and he would only require to drawl for a little while, until practice facilitated his command of the other sounds. Although producing the simple sound which we call the *e* of *berry*, or of the French words *de* or *que*, is a means of opening the glottis, which by stutterers is found very generally to answer, there are many cases in which other means are more suitable, as the intelligent preceptor soon discovers.—Were it possible to divide the nerves of the muscles which close the glottis, without at the same time destroying the faculty of producing voice, such an operation would be the most immediate and certain cure of stuttering ; and the loss of the faculty of closing the glottis would be of no moment.

" The view given above of the nature of stuttering and its cure, explains the following facts, which to many persons have hitherto appeared extraordinary. Stutterers often can sing well, and without the least interruption,—for the tune being continued, the glottis does not close. Many stutterers also can read poetry well, or any declamatory composition, in which the uninterrupted tone is almost as remarkable as in singing. The cause of stuttering being so simple as above described, one rule given and explained may, in certain cases, instantly cure the defect, however aggravated, as has been observed in not a few instances ; and this explains also why an ignorant pretender may occasionally succeed in curing, by giving a rule of which he knows not the reason, and which he cannot modify to the peculiarities of other cases. The same view of the subject explains why the speech of a stutterer has been correctly compared to the escape of liquid from a bottle with a long narrow neck, coming—" either as a hurried gush or not at all :" for when the glottis is once opened, and the stutterer feels that he has the power of utterance, he is glad to hurry out as many words as he can, before the interruption again occurs.

" Should the author's future experience enable him to simplify or render more complete the views of the nature and cure of stuttering, which he has given above, so as to facilitate the cure in every variety of case, he will not fail to publish his remarks."

ELEMENTS

OF

NATURAL PHILOSOPHY.

PART FOURTH.

DOCTRINES OF IMPONDERABLE SUBSTANCE, UNDER
THE HEADS OF HEAT, LIGHT, ELECTRICITY, AND
MAGNETISM.

To minds beginning this study, it may facilitate
the conception of a substance which is without
weight, or at least is imponderable by human art,
to consider the nature of *air*. Until lately men
were so imperfectly acquainted with the constitution of the universe around them, that a person
placed in an apartment which offered to view
nothing but the naked walls, would have said that
it was empty, meaning literally what he said; and
even when advertised that there was *air* in the
room, he would still have been far from possessing
a clear notion that it was full of aerial fluid, just
as an open vessel immersed in the sea is full of
water, and that if air were not allowed to escape
from it, even so small a body as an apple could not
be pressed into it additionally by less force than
fifty or sixty pounds. This truth however is now
clearly understood, and daily exemplified in easy

pneumatic experiments, and in no way more strikingly than by the recent adoption of the substance of air in place of feathers, as stuffing for beds and pillows. An air-tight bag or sack suspended by its lip in the air and held quite open by a hoop near its mouth, would appear empty, but if then firmly closed above the hoop, it would have imprisoned its fill of air, just as a bag similarly managed under water would imprison its fill of water ;—and while in some respects the air would be softer and locally more yielding than feathers, its entire mass would be much less compressible. Now this air, when weighed by means which modern science has furnished, is found in a cubic foot to contain somewhat more than an ounce, and by strongly pressing it, or by causing it to combine chemically with some other substance, we can reduce it to very small bulk, either with the form of a liquid or of a solid : proving how small a quantity of ponderable matter under certain circumstances will occupy great space. And common air is by no means the lightest known substance, which as powerfully resists the intrusion of other bodies where it exists. Hydrogen gas, for instance, of the same space-occupying force, weighs only a fourteenth part as much, and therefore a few drachms of it confined in a bag or bed as broad as the foundation of a house, would support a house or a cask as large as a house filled with water to a height of thirty feet, the gas itself being then eighty thousand times lighter than its bulk of gold ;—and if the pressure on it were diminished, it would readily expand to a volume a thousand times as great, and would still be exerting a considerable outward elasticity. Again, a mixture of oxygen and hydrogen gases, while uniting with explosive force to form water, dilates for the time, even under the great pressure of the atmosphere, to a bulk about twenty times greater than the gases have while separate.

The mind, pursuing the idea of such expansion or occupancy of space by a small quantity of matter, and reflecting on the wonderful divisibility of matter or minuteness of the ultimate atoms, as explained in Part I. of this work, might almost admit as a possible reality Newton's hypothetical illustration of that divisibility, *viz.* that even one ounce of substance uniformly distributed over the vast space in which our solar system exists, might leave no quarter of an inch without its particle. Now a fluid in any degree approaching in rarity to this, although it might press, resist, communicate motion, and have other influences in common with more ponderable matter, would have neither weight nor inertia discoverable by means at present known to man. While we are contemplating, then, or modifying the agencies of what causes the phenomena of heat and cold, of light and darkness, of electricity in its forms of thunder and lightning, of galvanism, or of magnetism, in a word, the most striking phenomena of nature, we may be dealing with matter of the subtile constitution now spoken of. And as in the terrestrial atmosphere there are at least two fluids present, *viz.* oxygen and nitrogen, of distinct nature, so in a more subtile ether filling all space, there may be various ingredients.

A majority of philosophers now incline to the opinion here sketched, that there is at least one such subtile fluid or ether occupying completely the space of the universe, and tending to uniform diffusion by reason of a strong mutual repulsion of its particles, which fluid pervades denser material substances somewhat as water pervades a sponge or a mass of sand, being attracted in a peculiar way by each substance, and which fluid may or may not have weight and inertia. They believe farther that the phenomena above alluded to, and which human art can exhibit with highest beauty, or with awful intensity, are produced by the motion or other affections of that fluid, as the sensation of *sound* in all its varieties is produced in the delicate structure of the ear by a certain motion in the air, or in any other body, having communication with the ear ; or as the sensation of *jar* is perceived by a hand held to one end of a log of wood when a blow is given to the other end. Some philosophers again suppose that the causes of the phenomena are material particles projected through space, somewhat as sand might be scattered by an explosion, and which particles are present only when the effects are apparent. Some combine these two hypotheses. And some hold all the phenomena of heat to be mere motions in the common matter of the bodies in which the heat exists.

We mention these hypotheses, not with the view of entering upon a minute examination of their respective merits, or even of asserting that any one of them is true, but merely to make the reader aware of the directions which inquirers' minds have taken in pursuing the investigation. To understand the subjects as far as men yet usefully understand them, and sufficiently for a vast number of most useful purposes, it is only necessary, as in other departments of science, to classify important phenomena, so that their nature and resemblances may be clearly perceived. When in treating of the human mind we speak of its *retaining an idea,* or being *depressed,* or being *heated with passion,* &c., we speak of subjects sufficiently definite, although we may have no hypothesis as to the intimate nature of the phenomena :—and in the same manner may we speak of the accumulation, radiation, or other affections of heat and light. We know nothing of the cause even of gravity, the grandest influence in nature, but we can calculate its effects with admirable precision.

PART FOURTH.

SECTION I.—ON HEAT.

ANALYSIS OF THE SECTION.

Heat (by some called Caloric) may be strikingly referred to as that which causes the difference between winter and summer, between tropical gardens and polar wastes. Its inferior degrees are denoted by the term COLD. *It cannot be exhibited apart, nor proved to have weight or inertia, and the change of its quantity in bodies is most conveniently estimated by the concomitant change of their bulk ; any substance so circumstanced as to allow this to be accurately measured constituting a* THERMOMETER.

Heat diffuses itself among neighbouring bodies until all have the same temperature, that is, until all similarly affect a thermometer. It spreads partly through their structure, or by conduction, as it is called, with a slow progress, different for each substance, and in fluids modified by the motion of their particles ; and it spreads partly also by being shot or radiated like light from one body to another, through transparent media or space, with readiness affected by the material and state of the giving and receiving surfaces.

Heat, by entering bodies, expands them, and through a range which includes, as three successive stages, the forms of SOLID, LIQUID, *and* AIR *or* GAS *; becoming thus in nature the grand antagonist and modifier of that attraction which holds corporeal particles together, and which, if acting alone, would reduce the whole material universe to one solid lifeless mass. Each particular substance, according to the nature, proximity, &c. of its ultimate particles, takes a certain quantity of heat (said to mark its capacity), to produce in it a given change of temperature or calorific tension ; undergoing expansion then in a degree proper to itself, and changing its form to liquid and air at points of temperature proper to itself ;—— the expansion in bodies generally increasing more rapidly than* *the temperature, because the cohesion of their particles lessens with increase of distance ; being remarkably greater therefore in liquids than in solids, and in airs than in liquids ; and the rate of expansion, moreover, being much quickened as the bodies approach their points of changing form to liquid or air, to produce which changes, a large quantity of heat enters them, but in the new arrangement of particles and increased volume of the mass, it becomes hidden from the thermometer, and is therefore called* LATENT HEAT. *For any given substance the changes of form happen so constantly at the same temperature, that they mark fixed points in the general scale of temperature, and enable us to regulate and compare thermometers.—Heat by expanding different substances unequally influences much their chemical combination.—Heat influences also the functions of vegetable and animal life.*

The great source of heat is the sun ; but electricity, combustion, and other chemical actions, condensation, friction, and the actions of life, are also excitants. *

" *Heat may be strikingly referred to as that which causes the difference between winter and summer, between the gardens of the equator and polar wastes.*" (See the Analysis, page 6.)

In the winter of climates, where the temperature is for a time below the freezing point of water, the earth with its waters is bound up in snow and ice, the trees and shrubs are leafless, appearing everywhere like withered skeletons, countless

* It is to be remarked here, that many phenomena in which heat plays an important part, have been already described in preceding chapters of this work ;—for instance, the action of the steam-engine, the phenomena of winds, many facts in meteorology, &c. under the head of Pneumatics. In a separate treatise on heat, these could not with propriety have been omitted ; but in a comprehensive system of science like the present, they find their fit place, where, being surrounded by subjects resembling them in more intricate particulars, they can be more concisely and clearly explained.

multitudes of living creatures, owing either to the bitter cold or deficiency of food, are perishing in the snows—nature seems dying or dead; but what a change when spring returns, that is, when heat returns! The earth is again uncovered and soft, the rivers flow, the lakes are again liquid mirrors, the warm showers come to foster vegetation, which soon covers the ground with beauty and plenty. Man, lately inactive, is recalled to many duties; his water-wheels are everywhere at work, his boats are again on the canals and streams, his busy fleets of industry are along the shores:— winged life in new multitudes fills the sky, finny life similarly fills the waters, and every spot of earth teems with vitality and joy. Many persons regard these changes of season as if they came like the successive positions of a turning wheel, of which one necessarily brings the next; not adverting that it is the single circumstance of change of temperature which does all. But if the colds of winter arrive too early, they unfailingly produce the wintry scene, and if warmth come before its time in spring, it expands the bud and the blossom, which a return of frost will surely destroy. A seed sown in an ice-house never awakens to life.

Again, as regards climates, the earthy matters forming the exterior of our globe, and therefore entering into the composition of soils, are not different for different latitudes,—at the equator, for instance, and near the poles. That the aspect of nature then in the two situations exhibits a contrast more striking still than between summer and winter, is owing merely to an inequality of temperature, which is permanent. Were it not for this, in both situations the same vegetables might grow, and the same animals might find their befitting support. But now, in the one, namely, where heat abounds, we see the magnificent scene of tropical fertility : the earth covered with luxuriant vegetation in endless lovely variety, and even the hard rocks festooned with green, perhaps with the vine, rich in its purple clusters. In the midst of this scene, animal existence is equally abundant, and many of the species are of surpassing beauty—the plumage of the birds is as brilliant as the gayest flowers. The warm air is perfume from the spice-beds, the sky and clouds are often dyed in tints as bright as freshest rainbow, and happy human inhabitants call the scene a paradise. Again, where heat is absent, we have the dreary spectacle of polar barrenness, namely, bare rock or mountain, instead of fertile field ; water every where hardened to solidity, no rain, nor cloud, nor dew, few motions but drifting snow ; vegetable life scarcely existing, and then only in sheltered places turned to the sun—and instead of the palms and other trees of India, whose single leaf is almost broad enough to cover a hut, there are bushes and trees, as the furze and fir, having what may be called hairs or bristles in the room of leaves. In the winter time, during which the sun is not seen for nearly six months, new horrors are added, *viz.* the darkness and dreadful silence, the cold benumbing all life, and even freezing mercury—a scene into which man

may penetrate from happier climes, but where he
can only leave his protecting ship and fires for
short periods, as he might issue from a diving-
bell at the bottom of the ocean.   That in these
now desolate regions, heat only is wanted to
make them like the most favoured countries of
the earth, is proved by the recent discoveries
under-ground of the remnant of animals and vegetables formerly inhabiting them, which now can
live only near the equator.   While winter then,
or the temporary absence of heat, may be called
the sleep of nature, the more permanent torpor
about the poles appears like its death ; and when
we further reflect, that heat is the great agent in
numberless important processes of chemistry and
domestic economy, and is the actuating principle
of the mighty steam-engine which now performs
half the work of society, how truly may heat, the
subject of our present chapter, be considered as
the life or soul of the universe !

" *Heat cannot be exhibited in a separate state, nor
 proved to have weight or inertia.*"   (Read the
 Analysis, page 6.)

Although heat is known to be abundant in the
sun-beam, and to radiate around from a blazing
fire, we cannot otherwise arrest or detect it in its
progress than by allowing it to enter, and remain
in some ponderable substance.   We know hot iron
or hot water or hot air, but nature nowhere presents to us, nor has art succeeded in shewing us
heat alone.

If we balance a quantity of ice in a delicate

weigh-beam, and then leave it to melt, the equilibrium will not be in the slightest degree disturbed. Or if we substitute for the ice, boiling water or red-hot iron, and leave this to cool, there will be no difference in the result. If we place a pound of mercury in one scale of the weigh-beam and a pound of water in the other, and then either heat or cool both through the same number of thermometric degrees, although about thirty times more heat (as will be explained below) enters or leaves the bulky water than the dense mercury, they will still remain equivalent weights.

Again, a sun-beam, with its intense light and heat, after being concentrated by a powerful lens or mirror, may be made to fall upon the scale of a most delicate balance, but will produce no depressing effect on the scale, as would follow if what constitutes the beam had the least forward motal inertia or momentum.

Such are the facts which have led certain inquirers to deny the material or separate existence of heat, and to hold that it is merely motion of one kind among the material particles of bodies generally, as sound is motion of another kind among the same particles. The following facts they consider to have the same bearing in the argument. Heat can be produced without limit by friction, as—when savages light their fires by rubbing together two pieces of wood—when Count Rumford made great quantities of water boil, by causing a blunt borer to rub against a mass of metal immersed in the water—when Sir

Humphrey Davy quickly melted pieces of ice by rubbing them against each other in a room cooled below the freezing point, &c. Intense heat is produced by the explosion of gunpowder or other fulminating mixture, yet it cannot be conceived to have existed in the small bulk of the powder before the explosion. Other inquirers, on the contrary, have deemed to be proofs of the separate materiality of heat such facts as now follow :—that it is radiated through the most perfect vacuum which we can produce, and even more readily than through air ;—that it radiates in the same place in all directions, without impediment from the crossing rays ;—that it becomes instantly sensible on the condensation of any material mass, as if then squeezed out from the mass; as when by compressing air suddenly, we inflame a match immersed in it, or when, on reducing the bulk of iron by hammering, we render it very hot, the warming being greater at the first blow (which most changes the bulk) than afterwards,—that when, on mixing bodies which combine so intimately as to occupy less space than when separate, there is a disengagement of heat proportioned to the diminution of volume :—that the laws of the spreading of heat in bodies do not resemble those of the spreading of sound, or of any other motion known to us :—and that as to the great and sudden extrication of heat by friction or explosion, it may be as truly a rush of the fluid to the part, as in the case of an electrical accumulation or discharge. These facts, moreover, they think square well with their assumption that the phenomena of heat are produced by an exceedingly subtile fluid or ether pervading the whole universe, and softening or melting or gasifying bodies, according to the quantity present in each, its own parts being strongly repulsive of each other, and seeking therefore widest and most equable diffusion.

" *The change of its quantity in bodies is most conveniently estimated by the concomitant change of their bulk, any substance so circumstanced as to allow this to be accurately measured, constituting a thermometer.*" (Read the Analysis, page 6.)

If we heat a wire it is lengthened; if we heat water in a full vessel, a part runs over; if we heat air in a bladder, the bladder is distended: in a word, if we heat any substance, its volume increases in some proportion to the increase of temperature,— and we may measure the increase of volume. The reasons why, in such investigations, a contrivance in which the expansion of mercury may be observed, *viz.* the mercurial thermometer, is commonly preferred to others, can only be fully understood by the mind which has considered the whole subject of heat; and we touch upon the matter here, only for the purpose of stating that a mercurial thermometer is a small bulb or bottle of glass filled with mercury, and having a long very narrow stalk or neck, in which the mercury rises when expanded by heat, or falls when heat is withdrawn; the stalk between the points at which the mercury stands in freezing and in boiling water, being divided into an arbitrary number of degrees, which division appear-ing on a scale applied to the stalk, is continued similarly above and below these points.

" *Heat diffuses itself among neighbouring bodies until all have acquired the same temperature ; that is to say, until all will similarly affect a thermo-meter.*" (See the Analysis, page 6.)

An iron bolt thrust in among burning coals soon becomes red-hot like them. If it be the heater of a tea-urn, it will, when afterwards placed amidst the water, part with its lately acquired heat to the water, until both are of the same tem-perature. Boiling-water, again, soon imparts heat to an egg placed in it, and a feverish head yields its heat to a bladder of cold-water or ice. A hun-dred objects enclosed in the same apartment, if tested after a time by the thermometer, will all indicate the same temperature.

" *The inferior degrees of heat are denoted by the term* COLD."

When the hand touches a body of higher tem-perature than itself, it receives heat according to the law now explained, and it experiences a pe-culiar sensation ; when it touches a body of lower temperature than itself, it gives out heat for a like reason, and experiences another and very different sensation. The two are called the sen-sations of *heat* and of *cold*. Now heat and cold, considered as existing in the bodies themselves, although thus appearing opposites, are really de-grees of the same object, *temperature*, contrasted by name for convenience sake, in reference to the particular temperature of the individuals speaking of them—just as any two nearest mile-stones on a road, although merely marking degrees of the same object, *distance*, might receive from persons living between them the opposite names of east and west, or of north and south. It is to be remarked, moreover, that the sensation of heat is producible also by a body colder than the hand, provided it be less cold than a body touched immediately before, or than the usual temperature; and the sensation of cold is producible under the opposite circumstances of touching a comparatively warm body, but which is less warm than something touched just before. This explains the remarkable fact that the same body may appear at the same time, and to the same person, both hot and cold. If a person transfer one hand to common spring-water from touching ice, that hand will deem the water very warm; while the other hand, transferred to it from a warm bath, would deem it very cold. For a like reason, a person from India, arriving in England in the spring, deems the air cold, while the inhabitants of the country are diminishing their clothing because the heat to them is becoming oppressive. Such facts shew how necessary it was for men to discover more correct thermometers than their bodily sensations.

" *Spreading partly through their structure, or by conduction, as it is called, with a progress proper to each substance.*" (Read the Analysis, page 6.)

If one end of a rod of iron be held in the fire,
a hand grasping the other end soon feels the heat
coming through it. Through a similar rod of
glass, the transmission is much slower, and through
one of wood it is slower still. The hand would
be burned by the iron before it felt warmth in the
wood, although the inner end were blazing.

On the fact that different substances are permeable to heat, or have the property of conducting it, in different degrees, depend many interesting phenomena in nature and in the arts:
hence it was important to ascertain the degrees
exactly, and to classify the substances. Various
methods for this purpose have been adopted. For
solids—similar rods of the different substances,
after being thinly coated with wax, have been
placed with their inferior extremities in hot oil,
and then the comparative distances to which in
a given time the wax was melted, furnished one
set of indications of the comparative conducting
powers :—or, equal lengths of the different bare
rods being left above the oil, and a small quantity
of explosive powder being placed on the top of
each, the comparative intervals of time elapsing
before the explosions gave another kind of measure :—or, equal balls of different substances, with
a central cavity in each to receive a thermometer,
being heated to the same degree and then suspended in the air to cool until the thermometer
fell to a given point, gave still another list. A
modification of the last method was adopted by
Count Rumford to ascertain the relative degrees
in which furs, feathers, and other materials used

for clothing, conduct heat, or which is the same thing, resist its passage. He covered the ball and stem of a thermometer with a certain thickness of the substance to be tried, by placing the thermometer in a larger bulb and stem of glass, and then filling the interval between them with the substance; and after heating this apparatus to a certain degree, by dipping it in liquid of the desired temperature, he surrounded it by ice, and marked the comparative times required to cool the thermometer a certain number of degrees. The figures following the names of some of the substances in the subjoined list, mark the number of seconds required respectively for cooling it 60°.

These experiments have shewn as a general rule, that density in a body favours the passage of heat through it. The best conductors are the metals, and then follow in succession diamond, glass, stones, earths, woods, &c., as here noted:

Metals—silver, copper, gold, iron, lead.
Diamond.
Glass.
Hard stones.
Porous earths.
Woods.
Fats or thick oils.
Snow.
Air 576
Sewing silk 917
Wood ashes....................... 927
Charcoal 937
Fine lint 1,038

Cotton 1,046
Lamp black 1,117
Wool 1,118
Raw silk 1,284
Beavers' fur 1,296
Eider down 1,305
Hares' fur........................ 1,315

Air appears near the middle of the preceding list, but if its particles are not allowed to move about among themselves, so as to *carry* heat from one part to another, it *conducts* (in the manner of solids) so slowly that Count Rumford doubted whether it conducted at all. It is probably the worst conductor known, that is, the substance which when at rest impedes the passage of heat the most. To this fact seems to be owing in a considerable degree the remarkable non-conducting quality of porous or spongy substances, as feathers, loose filamentous matter, powders, &c., which have much air in their structure, often adherent with a force of attraction which immersion in water, or even being placed in the vacuum of an air-pump, is insufficient to overcome.

While contemplating the facts recorded in the above table, one cannot but reflect how admirably adapted to their purposes the substances are which nature has provided as clothing for the inferior animals ;—and which man afterwards accommodates with such curious art to his peculiar wants. Animals required to be protected against the chills of night and the biting blasts of winter ; and some of them which dwell among eternal ice, could not have lived at all, but for a garment which might shut up within it nearly all the heat

which their vital functions produced. Now any
covering of a metallic or earthy or woody nature,
would have been far from sufficing ; but out of a
wondrous chemical union of carbon with the soft
ingredients of the atmosphere, those beautiful
textures are produced called fur and feather, so
greatly adorning while they completely protect
the wearers :—textures, moreover, which grow
from the bodies of the animals, in the exact
quantity that suits the climate and season, and
which are reproduced when by any accident they
are partially destroyed. In warm climates the
hairy coat of quadrupeds is comparatively short
and thin ; as in the elephant, the monkey, the
tropical sheep, &c. It is seen to thicken with
increasing latitude, furnishing the soft and abundant fleeces of the temperate zones ; and towards
the poles it is externally shaggy and coarse, as in
the arctic bear. In amphibious animals, which
have to resist the cold of water as well as of air,
the fur grows particularly defensive, as in the otter
and beaver. Birds, from having very warm blood,
required plenteous clothing, but required also to
have a smooth surface, that they might pass easily
through the air :—both objects are secured by the
beautiful structure of feathers,—so beautiful and
wonderful that writers on natural theology have
often particularized it as one of the most striking
exemplifications of creative wisdom. Feathers,
like fur, appear in kind and quantity suited to particular climates and seasons. The birds of cold
regions have covering almost as bulky as their
bodies, and if it be warm in those of them which

live only in air, in the water-fowl it is warmer still. These last have the interstices of the ordinary plumage filled up by the still more delicate structure called down, particularly on the breast, which in swimming first meets and divides the cold wave. There are animals with warm blood which yet live very constantly immersed in water, as the whale, seal, walrus, &c. Now neither hair nor feathers, however oiled, would have been a fit covering for them ; but kind nature has prepared an equal protection in the vast mass of fat or thick oil which surrounds their bodies—substances which are scarcely less useful to man than the furs and feathers of land animals.

While speaking of clothing, we may remark, that the bark of trees is also a structure very slowly permeable to heat, and securing therefore the temperature necessary to vegetable life.

And while we admire what nature has thus done for animals and vegetables, let us not overlook her scarcely less remarkable provision of ice and snow, as winter clothing for the lakes and rivers, for our fields and gardens. Ice, as a protection to water and its inhabitants, was considered in vol. i. in the explanation of why, although solid, it swims on water. We have now to remark that snow, which becomes as a pure white fleece to the earth, is a structure which resists the passage of heat nearly as much as feathers. It, of course, can defend only from colds below 32° or the freezing point ; but it does so most effectually, preserving the roots and seeds and tender plants during the severity of winter. When the green blade of wheat and the beautiful snow-drop flower appear in spring rising through the melting snow, they have recently owed an important shelter to their wintry mantle. Under deep snow, while the thermometer in the air may be far below zero, the temperature of the ground rarely remains below the freezing point. Now this temperature, to persons some time accustomed to it, is mild and even agreeable. It is much higher than what often prevails for long periods in the atmosphere of the centre and north of Europe. The Laplander, who during his long winter lives under ground, is glad to have additionally overhead a thick covering of snow. Among the hills of the west and north of Britain, during the storms of winter, a house or covering of snow frequently preserves the lives of travellers, and even of whole flocks of sheep, when the keen north wind catching them unprotected, would soon stretch them lifeless along the earth.

It is because earth conducts heat slowly, that the most intense frosts penetrate but a few inches into it, and that the temperature of the ground a few feet below its surface is nearly the same all the world over. In many mines, even although open to the air, the thermometer does not vary one degree in a twelvemonth. Thus also water in pipes two or three feet under ground does not freeze, although it may be frozen in all the smaller branches exposed above. Hence, again, springs never freeze, and therefore become remarkable features in a snow-covered country. The living water is seen issuing from the bowels

of the earth, and running often a considerable
way through fringes of green, before the gripe of
the frost arrests it; while around it, as is well
known to the sportsman, the snipes and wild duck
and other birds are wont to congregate. A spring
in a frozen pond or lake may cause the ice to be so
thin over the part where it issues, that a skaiter
arriving there will break through and be destroyed. The same spring-water which appears warm
in winter, is deemed cold in summer, because,
although always of the same heat, it is in summer
surrounded by warmer atmosphere and objects.
In proportion as buildings are massive, they acquire more of those qualities, which have now
been noticed, of our mother earth. Many of the
gothic halls and cathedrals are cool in summer
and warm in winter—as are also old-fashioned
houses or castles with thick walls and deep cel-
lars. Natural caves in the mountains or sea-
shores furnish other examples of a similar kind.

When in the arts it is desired to prevent the
passage of heat out of or into any body or situation, a screen or covering of a slow conducting substance is employed. Thus, to prevent the
heat of a smelting or other furnace from being
wasted, it is lined with fire-bricks, or is covered
with clay and sand, or sometimes with powdered
charcoal. A furnace so guarded may be touched
by the hand, even while containing within it
melted gold. To prevent the freezing of water
in pipes during the winter, by which occurrence the pipes would be burst, it is common to
cover them with straw-ropes, or coarse flannel,

or to enclose them in a larger outer pipe, with dry charcoal, or sawdust, or chaff, filling up the interval between. If a pipe, on the contrary, be for the conveyance of steam or other warm fluid, the heat is retained, and therefore saved by the very same means. Ice-houses are generally made with double walls, between which dry straw placed, or sawdust, or air, prevents the passage of heat. Pails for carrying ice in summer, or intended to serve as wine-coolers, are made on the same principle—*viz.* double vessels, with air or charcoal, filling the interval between them. A flannel covering keeps a man warm in winter—it is also the best means of keeping ice from melting in summer. Urns for hot water, tea-pots, coffee-pots, &c. are made with wooden or ivory handles, because, if metal were used, it would conduct the heat so readily that the hand could not bear to touch them.

It is because glass and earthenware are brittle, and do not allow ready passage to heat, that vessels made of them are so frequently broken by sudden change of temperature. On pouring boiling water into such a vessel, the internal part is much heated and expanded (as will be explained more fully in a subsequent page) before the external part has felt the influence, and this is hence riven or cracked by its connection with the internal. A chimney mirror is often broken by a lamp or candle placed on the marble shelf too near it. The glass cylinder of an electrical machine will sometimes be broken by placing it near the fire, so that one side is heated while the other side receives a cold current of air approaching the fire from a door or window. A red-hot rod of iron drawn along a pane of glass will divide it almost like a diamond knife. Even cast-iron, as backs of grates, iron pots, &c., although conducting readily, is often, owing to its brittleness, cracked by unequal heating or cooling, as from pouring water on it when hot. Pouring cold water into a heated glass will produce a similar effect. Hence glass vessels intended to be exposed to strong heats and sudden changes, as retorts for distillation, flasks for boiling liquids, &c., are made very thin, that the heat may pervade them almost instantly and with impunity.

There is a toy called a *Prince Rupert's Drop,* which well illustrates our present subject. It is a lump of glass let fall while fused into water, and thereby suddenly cooled and solidified on the outside before the internal part is changed; then as this at last hardens and would contract, it is kept extended by the arch of external crust, to which it coheres. Now if a portion of the neck of the lump be broken off, or if other violence be done, which jars its substance, the cohesion is destroyed, and the whole crumbles to dust with a kind of explosion. Any glass cooled suddenly when first made remains very brittle, for the reason now stated. What is called the *Bologna jar* is a very thick small bottle, thus prepared, which bursts by a grain of sand falling into it. The process of annealing, to render glassware more tough and durable, is merely the allowing it to cool very slowly by placing it in an oven, where the temperature is caused to fall gradually. The tempering of metals by sudden cooling seems to be a process having some relation to that of rendering glass hard and brittle.

It is the difference of conducting power in bodies which is the cause of a very common error made by persons in estimating the temperature of bodies by the touch. In a room without a fire all the articles of furniture soon acquire the same temperature ; but if in winter, a person with bare feet were to step from the carpet to the wooden floor, from this to the hearth-stone, and from the stone to the steel fender, his sensation would deem each of these in succession colder than the preceding. Now the truth being that all had the same temperature, only a temperature inferior to that of the living body, the best conductor, when in contact with the body, would carry off heat the fastest, and would therefore be deemed the coldest. Were a similar experiment made in a hot-house, or in India, while the temperature of every thing around were 98°, *viz.* that of the living body, then not the slightest difference would be felt in any of the substances : or lastly, were the experiment made in a room where by any means the general temperature were raised considerably above blood-heat, then the carpet would be deemed considerably the coolest instead of the warmest, and the other things would appear hotter in the same order in which they appeared colder in the winter room. Were a bunch of wool and a piece of iron exposed to the severest cold of Siberia, or of an artificial frigorific mixture, a man might touch the first with impunity (it would merely be felt as rather cold) ; but if he grasped the second, his hand would be frost-bitten and possibly destroyed : were the two substances, on the contrary, transferred to an oven, and heated as far as the wool would bear, he might again touch the wool with impunity (it would then be felt as a little hot), but the iron would burn his flesh. The author has entered a room where there was no fire, but where the temperature from hot air admitted was sufficiently high to boil the fish, &c. of which he afterwards partook at dinner ; and he breathed the air with very little uneasiness. He could bear to touch woollen cloth in this room, but no body more solid.

The foregoing considerations make manifest the error of supposing that there is a positive warmth in the materials of clothing. The thick cloak which guards a Spaniard against the cold of winter, is also in summer used by him as protection against the direct rays of the sun :—and while in England flannel is our warmest article of dress, yet we cannot more effectually preserve ice than by wrapping the vessel containing it in many folds of softest flannel.

In every case where a substance of different temperature from the living body touches it, a thin surface of the substance immediately shares the heat of the bodily part touched—the hand generally ; and while in a good conductor, the heat so received quickly passes inwards, or away from the surface, leaving this in a state to absorb more, in the tardy conductor the heat first received tarries at the surface, which consequently soon acquires nearly the same temperature as the hand, and therefore, however cold the interior of the substance may be, it does not cause the sensation of cold. The hand on a good conductor has to warm it deeply, a slow conductor it warms only superficially. The following cases farther illustrate the same principle. If the ends of an iron poker, and of a piece of wood of the same size, be wrapped in paper and then thrust into a fire, the paper on the wood will begin to burn immediately, while that on the metal will long resist : ─or if pieces of paper be laid on a wooden plank and on a plate of steel, and then a burning coal be placed on each, the paper on the wood will begin to burn long before that on the plate. The explanation is, that the paper in contact with the good conductor loses to this so rapidly the heat received from the coal, that it remains at too low a temperature to inflame, and will even cool to blackness the touching part of the coal ; while on the tardy conductor the paper becomes almost immediately as hot as the coal. It is because water exposed to the air cannot be heated beyond 212°, that it may be made to boil in an egg-shell or a vessel made of paper, held over a lamp, without the containing substance being destroyed ; but as soon as it is dried up, the paper will burn and the shell will be calcined, as the solder of a common tinned kettle melts under the same circumstances. The reason why the hand judges a cold liquid to be so much colder than a solid of the same temperature is, that from the mobility of the liquid particles among themselves, those in contact with the hand are constantly changing. The impression produced on the hand by very cold mercury is almost insufferable, because mercury is both a ready conductor and a liquid. Again, if a finger held motionless in water feel cold, it will feel colder still when moved about ; and a man in the air of a calm frosty morning does not experience a sensation nearly so sharp as if with the same temperature there be wind. A finger held up in the wind discovers the direction in which the wind blows by the greater cold felt on one side, the effect being still more remarkable, if the finger is wetted. If a person in a room with a thermometer, were with a fan or bellows to blow the air against it, he would not thereby lower it, because it had already the same temperature as the air, yet the air blown against his own body would appear colder than when at rest, because, being colder than his body, the motion would supply heat-absorbing particles more quickly. In like manner, if a fan or bellows were used against a thermometer hanging in a furnace or hot-house, the thermometer would suffer no change, but the air moved by them against a person would be distressingly hot, like the blasting sirocco of the sandy deserts of Africa. If two similar pieces of ice be placed in a room somewhat warmer than ice, one of them may be made to melt much sooner than the other, by blowing on it with a bellows. The reason may here be readily comprehended why a person suffering what is called a cold in the head, or catarrh from the eyes and nose, experiences so much more relief on applying to the face a handkerchief of linen or cambric than one of cotton :—it is, that the former by *conducting* readily absorbs the heat and diminishes the inflammation, while the latter, by refusing to give passage to the heat, increases the temperature and the distress. Popular prejudice has held that there was a poison in cotton.

" *Heat spreading in fluids chiefly by the motion of their particles.*" (Read the Analysis, page 6.)

Owing to the mobility among themselves of fluid particles, heat entering a fluid anywhere below the surface, by dilating and rendering specifically lighter the portion heated, allows the denser fluid around to sink down and force up the rarer ; and the continued currents so established, diffuse the heat through the mass much more quickly than heat spreads by conduction in any solid.

Count Rumford's experiments led him at first to conclude that liquids, but for this carrying process, by the particles changing their place, were absolutely impassable to heat. A piece of ice will lie very long at the bottom of water which is made to boil at the top by the contact of any hot body ; and when it at last melts, Count R. believed that it did so entirely from the heat which passed downwards through the sides of the vessel containing the water. But an ingenious experiment by Dr. Murray decided the question differently. He made a vessel of ice, which of course could not carry downwards any heat greater than 32°, as ice melts at that degree; and having put into it a quantity of oil at 32°, the bulb of a thermometer being a quarter of an inch under the surface of the oil, he placed a cup of boiling water in contact with the surface of the oil :—in a minute and a-half the thermometer rose nearly a degree, and in seven minutes it rose five degrees, beyond which it did not go. The heat then must have passed downwards through the liquid, proving a conducting power ;—unless indeed it passed by radiation, as explained in a subsequent page.

The internal currents or circulation produced by heat in fluid masses, and of which there are so many important instances in nature, were more fitly explained in the chapter on *hydrostatics* and *pneumatics ;* we shall here therefore allude to them very shortly.

Perhaps the best experimental illustration of the subject is the placing a tall glass jar, filled with water in which small pieces of amber are floating to shew its movements, first in a warm bath, and then in one which is cold. In the first case, the water and amber near the outside of the jar where they are heated, will exhibit a rapid upward current, while in the centre of the jar they will form an opposite or downward current. By afterwards placing the jar in a cold bath, the direction of the currents will be reversed.

It is, as stated in the former volume, this heating and dilatation of the fluid air over a tropical island while acted upon during the middle of the day by the powerful rays of the sun, that allows the colder and heavier air from the face of the ocean around to press inwards upon it and force it upwards in the atmosphere—the cold current forming the delightful sea-breeze of the climate. And it is the general heating of the air over the whole equatorial belt of the earth, which, rendering it specifically lighter than the air nearer the poles, allows this to assume the form of cool trade winds, constantly blowing towards the sun's path, and pressing upwards the hot air, which then spreads away on the top of the atmosphere towards the poles, to mitigate the severity of the northern and southern cold. In the watery ocean also there is a circulatory motion of the same kind, although less in degree, tending to distribute heat and equalize temperature, and contributing to produce some of the great sea currents known to mariners.

The vertical currents produced by heat, in the ocean and in great masses of water generally, preserve in and over them a comparatively uniform temperate freshness, while the rocks and soil around may be either parched under a burning sun, or bound up in cold many degrees below freezing. A keen frost chills, and soon hardens in its icy grasp the surface of the ground; but of water similarly exposed the part first cooled descends to the bottom by its increased density, and forces up a warmer water to take its place; this in its turn is cooled and descends, and a continued circulation is established, so that the surface cannot become ice until the whole mass, of whatever depth, has been cooled down to its greatest density. The very deep sea hence is not frozen in the coldest climates, and in temperate climates the severest winter does not freeze even a deep lake. During this intestine movement in the water, that which ascends to the surface to be cooled, by losing one degree of its heat, warms nearly 500 times its bulk of air one degree, and thus tempers remarkably the air passing over it. Hence places in the vicinity of the sea and of lakes are warmer in winter than places further inland, although nearer the equator. England is much warmer in winter than central Germany, which lies south of England, and the coasts of Scotland and the north of Ireland are warmer than London :—snow never lies long upon these coasts. As continental or inland countries have thus in winter an extreme of cold, so have they in summer an extreme of heat. Water admits the rays of the sun, and absorbs the heat into the whole thickness of its mass (it will be explained afterwards that solar heat penetrates transparent fluids as light does), while earth retains all near its surface, and is therefore heated to excess.

The ventilation of our dwellings and halls of assembly (as explained in vol. i.) is owing to the motion produced by the changed specific gravity of air when heated. The air which is within the house becomes warmer than the external air, and this then presses in at every opening or crevice to displace it. The ventilation of the person by the slow passage of air through the texture of our clothing is a phenomenon of the same kind; and thicker clothing acts chiefly by diminishing the passage. Hence an oiled-silk or other air-tight covering laid on a bed, has greater influence in preserving warmth than an additional blanket or more. From the part of bed-clothes immediately over the person there is a constant outward oozing of warm air, and there is an oozing inward of cold air in lower situations around. In many persons the circulation of the blood is so feeble that in winter, they have great difficulty in keeping their feet warm, even in bed, unless with the assistance of a bottle of hot water or some such means, and in consequence they often pass sleepless nights, and suffer in their general health. At the suggestion of the author, in such cases, a long flexible tube has been used,—as of spiral wire covered with leather or oiled silk, by which persons can send down to their lower extremities their hot breath, and thus supply to them effectually a natural animal warmth, as in a cold day they warm their hands by blowing upon them through their gloves.

The power of fluids to diffuse heat being due thus to their power of *carrying*, and not of *conducting* it, the consequence should follow, that any circumstance which impedes the internal motion of the fluid particles, should diminish the diffusing power. Accordingly we find, that fluids in general transfer heat less readily in proportion as they are more viscid. Water, for instance, transfers less quickly than spirits; oil than water; molasses or syrup than oil: and water thickened by starch dissolved in it, or which has its internal motion impeded by feathers or thread immersed in it, less quickly than where it is pure and at liberty. Cooling being merely a motion the reverse of heating, it is influenced by the same law. Hence the reason why soups, pies, puddings, and all semifluid masses, retain their heat so long—so much longer than equal bulks of mere fluid. The same law affords explanation of the facts, that very porous masses and powders, as charcoal, metal filings, sawdust, sand, &c. conduct heat more slowly than denser masses,—their interstices being filled with air, which scarcely *conducts* heat, and which, by the structure of the substance, has no freedom of motion or circulation by which it might *carry* the heat.

" *Heat spreads also, partly, by being shot or radiated like light, from one body to another, through transparent media or space, with readiness affected by the material and the state of the giving and receiving surfaces.*" (Read the Analysis, page 6.)

If a heated ball of metal be suspended in the air, a hand brought in any direction near to it will experience the sensation of heat : and beneath it the sensation will be as strong as on the side, although the heat has to shoot down through an opposing current of air approaching the heated ball, to rise from it, as explained in a preceding section. A delicate thermometer substituted for the hand will equally detect the spreading heat, and if held at different distances, will prove it to diminish in the same ratio as light diminishes in spreading from any luminous centre, *viz.* to be only a fourth part as intense at a double distance, and in a corresponding proportion for other distances. If the heated body be enclosed in a vacuum, a thermometer placed near it will still be affected in the same manner. If a screen be interposed between the body and the thermometer, the latter will not be affected at all, proving the heat to spread in straight lines. Heat when diffusing itself in this way, to distinguish it from heat passing by contact or communication, as described in the last section, is called *radiant* heat or caloric ; that is to say, spreading in rays all round its source as light spreads.

Radiant heat resembles light yet in other respects. It as rapidly permeates certain transparent substances, and its course suffers in them a degree of the bending, termed refraction by opticians. It is reflected from many kinds of polished surfaces, just as light is reflected from a common mirror ; and many such surfaces directed to one centre (as when Archimedes made the sun his assistant to burn the Roman ships) or a single concave surface, having its own centre or focus, will concentrate heat just as light. Its motion in the sun-beam is so rapid, as for any distance at which men can try the experiment to appear instantaneous ; and the rays of heat from hot iron or burning charcoal concentrated at great distances by suitable mirrors, affect a thermometer as quickly as the heat of the setting sun reflected from a distant window. Although light and heat are united in the sun's ray, they are still separable by our

D 2

glass prisms or lenses ; and the focus of heat behind a burning glass is not precisely the focus of light. Heat in radiating through air does not warm it, and is not affected by winds or any other motion of the air.——These resemblances in the phenomena of light and heat have by some inquirers been held to prove that the two classes of appearances are only different modifications of action in the same subtle substance or ether.

The diffusion of heat by radiation, as it takes place in an instant to any distance, and begins whenever there is any inequality of temperature between bodies exposed to each other, would produce instant balance of temperature throughout nature, but that heat leaves and enters bodies with readiness depending on their internal conducting powers, and on the condition of their surfaces. A black stone-ware teapot, for instance, will radiate away 100 degrees of its heat in the same time that a pot of polished metal will radiate 12 degrees.

Professor Leslie was the first to see the importance of investigating this subject, and he had the merit of contriving well-adapted means, and of detecting many of the important facts. As common thermometers are not sufficiently delicate to determine very sudden changes of temperature, where the influence is so slight as in many cases of radiant heat, he used the beautiful *differential thermometer* contrived by himself, in conjunction with concave mirrors to concentrate the heat and accumulate its energy. Then taking as his heated body a cubical tin vessel filled with boiling water, and covering it successively with plates or layers of different substances, and with different colours; and exposing the thermometer to it under all the changes, he noted the number of degrees which the thermometer rose—as seen in the table which here follows, and thus ascertained the radiating power of each sort of covering.

Lamp black .... 100°
Writing paper 98
Crown glass 90
Ice 87
Isinglass........................... 75
Tarnished lead 45
Clean lead 19
Iron polished.................... 15
Tin plate 12
Gold, silver, and copper 12

He next reversed the experiments by using his hot-water vessel always in the same state, and covering the thermometer bulb with the different substances and colours, and thus he ascertained that their comparative *absorbing* powers were very nearly proportioned to their *radiating* powers: lamp-black, for instance, absorbed or was heated 100°, while the polished metals absorbed, or were heated only 12°, and so for the others. And, lastly, the absorbing powers being likewise an indication of the opposite or *reflecting* powers (for if a body absorb any given proportion of the heat which falls on it, it must reflect the remainder), he had at the same time ascertained the reflective or mirror powers of the bodies, and therefore all the important points respecting radiant heat in its relation to the substances between which it passes.

It seems paradoxical that, by putting a clothing of a thin cotton or woollen fabric upon the polished tin vessel, the heat should be received by it or dissipated from it much sooner than if the vessel were naked, but such is the fact. And metal with a scratched or roughened surface radiates or receives much more rapidly than if polished.

The property of absorbing heat depends much upon the colour of the substance, and as a general rule the dark colours, *viz.* those which absorb most light, absorb also most heat. Dr. Franklin laid pieces of cloth of different colours on snow, and during a given period in which the sun was shining on them, he noted this in the different depths to which, by melting the snow which was under them, they sunk. Hence appears the importance of having a white dress in summer, that by it, with the sun's light, the heat also may be repelled ; and a white dress in winter is good, because it radiates little. Polar animals have generally white furs. White horses are both less heated in the sun, and less chilled in winter, than those of darker hues.

The rate of cooling in bodies must be influenced by all the particulars noted above, *viz.* substance, surface, colour, and by the excess of heat in the cooling body as compared with those around it.

The concentrating apparatus used for experiments on the radiation of heat consists of two concave tin mirrors, here represented at *a* and *b*, so formed and placed in relation to each other, that all the rays of light or heat

issuing from the focus of one, as at *c*, shall be collected in the focus of the other, *d*. A stand under one focus *c* is intended to support the body giving out or receiving heat, and a stand under the other *d* is meant to support the thermometer. For farther explanation of the action of such mirrors, we may refer to what was said of the concentration of sound in the section on *Acoustics*, or to what follows in the section on *Optics*, on the concentration of light. The general rationale of such facts is, that heat, light, sound, an elastic ball, &c. reflected from any point of a surface, returns, if it fall perpendicularly to that point, in the same line by which it approached; but if it fall obliquely, or on one side of the perpendicular, it returns in a line deviating as much on the other side. Now the surfaces of concave mirrors are so formed, that every ray issuing from the focus shall, when reflected, become parallel to every other ray—as represented by the dotted lines in the figure; and it is the property of a similar mirror receiving parallel rays to make them all meet in its focus :—thus any influence radiating from *c* will be again collected at *d*. The purpose and effect of such mirrors in experiments on heat, is merely to concentrate feeble influences, so that they may be more accurately estimated. To shew their effect and mode of action, they may be placed exactly facing each other at any convenient distance, and then a hot body of any kind, as a ball of metal or a canister of boiling water, being left in one focus while a thermometer stands in the other, the thermometer will instantly rise ; although, if left in any intermediate situation nearer to the hot body, and therefore not in the focus, it will not be affected. If burning charcoal be placed in one focus and a readily combustible substance in the other, the latter will be set fire to at the distance of thirty feet or more.

If in one focus of the mirror apparatus described above, there be placed, instead of the canister of hot water, a piece of ice, the thermometer in the other focus immediately falls. This has been called the radiation of cold, and persons were at one time disposed to think that it proved cold to have a positive existence distinct from heat. The case, however, is merely that the thermometer happens then to be the hotter body in one focus of the mirrors, placed in relation with a colder body, the ice, in the other, and consequently by the law of equable diffusion, it must share its heat with the ice, and will fall. The mirrors in any case have merely the effect, by preventing the spreading and dissipation of the radiant heat from either focus except towards the other, of making two distant bodies act upon each other as if they were very near. All the heat that seeks to radiate from the thermometer d in the direction of the large surface of the mirror *b*, if not met by an equal tension or force of temperature in the other mirror or focus to which they are directed at *a* and *c*, will radiate away to *c*, and become deficient at *d*. Some inquirers have believed that heat was constantly radiating in exchange from substance to substance (as light radiates constantly between opposed bodies), only more copiously from one side, if the temperature of that was higher : others have held that the movement only took place when the balance of temperature was destroyed; and this is the simplest view.

There is a remarkable difference in one respect between the heat of the sun and that radiated from any other source, *viz.* that the first passes through air, glass, water, and transparent bodies generally, very readily, while the latter, although not obstructed by air, is almost totally intercepted or absorbed in passing through any of the other substances named. In our drawing-rooms it is common to have ornamented glass fire-screens, which, while they allow the light to pass, defend the face from the heat : but all persons know that the heat of the sun-beams, as well as their light, enters our green-houses through the glass which covers them. A glass screen interposed between the concave mirrors in the apparatus above described, destroys almost entirely the effect of the heated body in one focus, on the thermometer in the other, and the trifling effect really produced has appeared to some to be owing to the heat first absorbed by the screen on one of its sides, and then radiated from the other. This conclusion seemed to be supported by the fact that screens of metal or of glass, covered with lamp black, paper, &c., allow transmission nearly in proportion to their several absorptive and radiant powers. More careful experiments, however, have been held to prove that a small portion of the heat is suddenly radiated through the glass. A glass mirror reflects the light of a fire, but at first retains all the heat, and only radiates it afterwards as a hot body.

The doctrines of radiant heat make us aware of the importance of having vessels of polished metal for containing liquids or any thing which we desire to keep warm ; hence tea and coffee-pots, dishes for soup, &c. should be polished. As a black earthen teapot loses heat by radiation nearly in proportion to the number 100, while one of silver or other polished metal loses only as 12, there will be a corresponding difference in their aptitude for extracting the virtues of any substance infused in them. Pipes for the conveyance of steam or hot air, if left naked, should be of polished metal ; but after arriving at the place where they have to give out their heat, their surface should be blackened and rough. A coat of polished mail is not a cold covering. A mirror intended to reflect heat should be of highly polished metal : and such is the interior of a screen placed behind roasting meat. A fireman's mask is usually covered with tin foil. It is of advantage that the bottom of a tea-kettle or other cooking vessel be black, because the bottom has to absorb heat, but the top should be polished because it has to confine.

The interesting phenomenon of dew was not at all understood until lately, since the laws of radiant heat have been investigated. At sun-rise in particular states of the sky, every blade of grass and leaflet is found, not wetted, as if by a shower, but studded with a row of distinct globules most transparent and beautiful, bending it down by their weight, and falling like pearls when the blade is shaken. These are formed in the course of the night by a gradual deposition on bodies rendered by radiation colder than the air around them, of the moisture which rises invisibly from water surfaces into the air during the heat of the day. In a clear night the objects on the surface of of the earth radiate heat upwards through the air which impedes not, while there is nothing nearer than the stars to return the radiation; they consequently soon become colder, and if the air around has its usual load of moisture, part of this will be deposited on them, exactly as the invisible moisture in the air of a room is deposited on a cold bottle of wine when first brought from the cellar. Air itself seems not to lose heat by radiation. A thermometer placed upon the earth any time after sun-set until sun-rise next morning, generally stands considerably lower than another suspended in the air a few feet above it; owing to the radiation of heat upwards from the earth, while the air remains nearly in the same state. During the day, while the sun shines, the earth is much warmer than the air. The reason why the dew falls, or forms so much more

copiously upon the soft spongy surface of leaves
and flowers, where it is wanted, than on the hard
surface of stones and sand, where it would be of
no use, is the difference of their radiating powers.
There is no state of the atmosphere in which artificial dew may not be made to form on a body,
by sufficiently cooling it, and the degree of heat
at which it begins to appear is called the dew
point, and is an important particular in the me-
teorological report of the day. In cloudy nights
heat is radiated back from the clouds, and the
earth below not being so much cooled, the dew is
scanty or deficient. On the contrary, when un-
informed persons would least expect the dew, *viz.*
in warm very clear nights, and perhaps when the
beautiful moon invites to walking, and music adds
its charm, as in some of the evenings of autumn
with the harvest moon and harvest occupations—
then is the dew more abundant, and the danger
greater to delicate persons of taking harm by
walking among the grass.

" *Heat by entering bodies expands them, and
through a range which includes as three successive
stages the forms of solid, liquid, and air, or gas ;
becoming thus, in nature, the grand antagonist
and modifier of the effects of that attraction which
holds corporeal atoms together, and which, if acting
alone, would reduce the whole material universe
to one solid lifeless mass.* (Read the Analysis,
page 6.)

If an experimenter take a body which is as free
from heat as man can procure a body—a bar of

solid mercury, for instance, as it exists in a polar winter ; and if he then gradually heat such body, to whatever extent, it will acquire an increase of bulk with every increase of temperature : first, there will be simple enlargement or expansion in every direction ; then the mass will in addition be softened ; then it will be melted or fused, that is to say, in the case supposed, the solid bar will be reduced to the state of liquid mercury, with the cohesive attraction of the atoms nearly overcome ; if the mass be still farther heated, the atoms will be repelled from each other to much greater distances, constituting then a very elastic fluid called an air or gas, many hundred times more bulky than the same matter in the solid or liquid state, and capable of forcibly distending an appropriate vessel as common air distends a bladder ; susceptible, moreover, of dilating indefinitely farther, by farther additions of heat, or by diminution of the atmospheric pressure, against which it had to rise during its formation. A subsequent removal of the heat will cause a corresponding progress of contraction, and the various conditions or forms of the substance above enumerated, will be reproduced in a reverse order, until the solid mass again appear.

What is thus true of mercury is proved by modern chemical art to be true also of all the ponderable elements of our globe, and of many of the combinations of these elements,—as water, for instance, familiarly known in its three forms of *ice*, *water*, and *steam ;* although compound substances generally, by great changes

of temperature, are decomposed into their elements.

A student might at first have difficulty in believing that the beautiful variety of solid, liquid,
and air found among natural bodies, could depend
upon the quantities of heat in them, because
these forms are all seen existing at the same temperature ; but he afterwards learns that each substance has its peculiar relation or affinity to heat,
and that hence, while at the medium temperature
of the earth, some bodies contain so little as to
be solids—like the metals, stones, earths, &c. ;
others have enough to be liquids—as mercury,
water, oils, &c. ; and others enough to be airs—
as oxygen, nitrogen, hydrogen, &c. Men, until
better informed, are prone to deem the states in
which bodies are most frequently observed, to be
the natural states of such bodies ; and the Indian
king but reasoned in a usual way, who held the
Dutch navigators, newly arrived on his shores, to
be gross impostors, when they said that in their
country, at one time of the year, water became
so hard that they could walk upon it, and drive
their carriages upon it, and shape it into solid
blocks. All persons err like this king, who in
thinking of the different substances known to them,
regard their accidental state as to the cohesion
of particles—which state is really dependent on
the temperature of the bodies, and therefore
on the particular planet or situation on the planet
where they are found, to be in them an essential
natural character. As well might a person who
had never seen silk, but as a delicate gauze or

satin enveloping some lovely human form, refuse to recognize it in the unsightly coil of the worm which produces it.

The degrees in a general scale of temperature at which the most important substances in nature change their states from solid to liquid, or from liquid to air, will be noted in a future page. Here we have only to remark, that the differences are very great. Mercury melts at about 80° below the melting point of ice, and porcelain at about 30,000° above. There are some substances which require so high a temperature for their fusion or for their conversion into gas, that human art has difficulty, or even finds it impossible, to produce the changes by simple concentration of heat; but all such are readily soluble in some other substance, possessing already the form of liquid or air : as when gold and platinum are dissolved in nitro-muriatic acid—flint in the fluoric acid,—carbon in hydrogen gas. Now many persons may not have reflected that the dissolving a solid in any fluid menstruum is merely another mode of melting it by heat ; yet this is the truth, for the menstruum is itself fluid, only because of the much heat which it contains, and in dissolving the more obdurate substance, it does so merely because its attraction for the substance brings the particles into union with the heat which already exists in itself. Heat then is the only and universal solvent. Its influence is interestingly seen in the fact, that a fluid when heated can dissolve much more of a solid than when cold. Water while hot keeps dissolved twice as much of many

salts as it can when its temperature has fallen.
—There are again in nature many substances
having such an affinity for heat, that until
lately they have only been known as airs; and
even in the present advanced state of art, they
cannot by any degree of mere cooling be reduced
to the liquid or solid form; yet all such, when
pressure is added to the cooling, or when the chemical attraction for them of some other substances
which already exist in the liquid or solid state, is
made to co-operate, may be reduced. An instance
is afforded by oxygen, when made part of a liquid
acid, or of a solid ore.

Of solids, some on receiving heat become soft
before they are liquefied, as pitch, glue, iron, &c.;
others change completely at once, as ice in becoming water; and some pass at once to the state
of air, without therefore having assumed at all
the intermediate state of liquid—they are *sub-
limed*, as it is called, and on cooling again may be
caught in a powdery state, as seen in that form of
sulphur, or of benzoin, termed the *flower* of the
substance. Of the latter class also are camphor,
arsenic, corrosive sublimate, and the substance
called iodine, which last, from the state of rich
ruby crystals, on being heated becomes at once a
dense transparent gas of the same hue, and in
cooling resumes its crystalline form.

The reader having arrived at this place, may
peruse again with advantage five pages of vol. i.
(between pages 20 and 30 in the different editions)
which treat of the influence of heat on the *constitution of masses*.

" *Each particular substance, according to the nature, proximity, &c. of its ultimate particles, takes a certain quantity of heat (said to mark its* CAPACITY) *to produce in it a given change of temperature or calorific tension.*" (Read the Analysis, page 6.)

A pound of water, for instance, to raise its temperature one degree, takes thirty times as much heat as a pound of mercury. This may be proved in various ways. First, if the heat be derived from any uniform source, the water must remain exposed to it thirty times as long as the mercury. Second, if both substances, after being equally heated, be placed in ice until cooled to the freezing point, the heat which escapes again from the water will melt thirty times as much ice as that which escapes from the mercury. Third, when a pound of hot water is mixed with a pound of cold mercury, instead of the two becoming of a middle temperature, as is the case when equal quantities of hot and cold water are mixed—every degree of heat lost by the one becoming just a degree gained by the other—the pound of hot water, by giving up one degree to the pound of cold mercury, raises the temperature of the latter thirty degrees; and in the same proportion for other differences :—or on reversing the experiment, a pound of hot mercury will be cooled thirty degrees by warming a pound of water one degree.

Now each particular substance in nature, just as water or mercury, has its peculiar capacity for heat; and experiments made by the modes of mixture and of melting ice above described have

E

led to the construction of tables which exhibit the relations. The following short table is an abstract of these, shewing the comparative capacities of equal weights of some common substances. Water, for reasons of convenience, has been chosen as the standard of comparison. It appears, then, that a pound of hydrogen gas takes about twenty times more heat to produce in it a given change of temperature than a pound of water, while a pound of gold takes about twenty times less, and therefore four hundred times less than the hydrogen. The figures in the table, by marking the comparative capacities for heat of various substances, necessarily indicate also the comparative quantites of ice which would be melted by equal weights of the substances in cooling through an equal number of degrees. A pound of water, the standard, must cool 140 degrees, that is, must give up 140 degrees of its heat to melt one pound of ice.

Gases.

Hydrogen	$21\frac{1}{2}$
Atmospheric air	$1\frac{3}{4}$
Carbonic acid gas	$1\frac{3}{5}$
Common steam	$1\frac{1}{2}$

Liquids.

Solution of carbonate of ammonia	2
Alcohol	$1\frac{1}{10}$
Water	1
Milk	1
Olive oil	$\frac{3}{4}$
Linseed oil	$\frac{1}{2}$

Sulphuric acid .... $\frac{1}{3}$

Quicksilver .... $\frac{1}{30}$

Solids.

Ice............................... $\frac{9}{10}$

Wheat .... $\frac{1}{2}$

Charcoal........................ $\frac{1}{3}$

Chalk .... $\frac{1}{4}$

Glass .... $\frac{1}{5}$

Iron .... $\frac{1}{8}$

Zinc .... $\frac{1}{10}$

Gold .... $\frac{1}{20}$

We may remark here that some late researches, by another mode of trial, make the capacity of air to be only a quarter that of water, although in the preceding table it appears to be one and three-quarters. Now as the other aeriform fluids have been compared with water through the medium of atmospheric air, if there be an error with respect to this, it must run through all the figures noting the capacity of other aeriform substances.

If we seek a reason or reasons why there should be among bodies the differences of capacity here stated, the circumstances chiefly calling attention are the following. First, equal weights of the various substances have very different bulks or volumes, and therefore have different room in which the heat may lodge. Mercury, for instance, is only one-fourteenth part as bulky as water. That the bulk, however, is not the only influencing circumstance appears in the fact, that mercury while having one-fourteenth of the bulk of water, has only one-thirtieth of the capacity. Second, in equal bulks of different substances, the space may be more completely occupied by the particles of one than of another—as by the particles of mercury than by those of water. But as the facts are not fully accounted for, even by both of these circumstances, we must seek explanation in a third, *viz.* a difference in the ultimate particles of bodies affecting their relations to heat.

First. The influence of bulk or volume, in determining the capacity for heat, is proved by the facts stated in the preceding table, and by many others. In the table, for instance, it is seen that hydrogen and the gases generally, with their great comparative bulk, have also great capacity; that liquids have less capacity than gases; that solids have less than liquids—but the capacity, as already stated, is not in strict proportion to bulk; for hydrogen, which is many thousand times more bulky than an equal weight of water, has only twenty-one times the capacity. Again, if any body whatever be suddenly compressed into less bulk, heat will issue from it as if squeezed out. Thus iron or other metal suddenly condensed by the heavy blow of a hammer, is thereby rendered hotter, and the expelled heat will gradually spread from it. Because water and spirit, on being mixed, occupy less space than when separate, there is from the mixture a corresponding discharge of heat. But the truth is most remarkably exemplified in airs or gases, owing to their great range of elasticity. They may be condensed or dilated a hundred-fold or more, and there will be a simultaneous concentration or diffusion of their heat, that is to say, the production, in the space occupied by them, of intense heat or cold. The heat of air just condensed, or the cold of that which has just expanded, is much greater than even the most delicate thermometer can indicate, for there is so little heat altogether even in a considerable volume of air, that the mass of a mercurial thermometer, although absorbing a great part of it, would be little affected. The extent, however, of the change of temperature is seen in the facts, that by the sudden condensation of air we may inflame tinder immersed in it, and by allowing air suddenly to expand, we may convert any watery vapour diffused through it into ice or snow. Nay, air, containing carbon in perfect solution, as is true of the common coal gas, if first condensed to expel heat, and then allowed suddenly to expand, will be so cooled that the carbon will be separated like a black cloud, as snow is separated in the case before described. The cold which separates or freezes carbon from a gas holding it in solution, is perhaps the most intense which art can produce. It might be expected that air suddenly compressed into half its previous volume, should become just twice as hot as before, or if suddenly dilated to double volume, should be only half as hot, thus enabling us to ascertain the whole quantity of heat contained in it; but the facts are not so ; the temperature changes, near the middle degrees of the scale at least, much less than the density. Air in doubling its volume from a common density, becomes colder only by about 50° of Fahrenheit's thermometer.

The different capacity for heat of air in different states of dilatation, produces effects of great
importance in nature as well as in the arts—
thus,

On the surface of the earth near the sea-shore,
the air of the atmosphere has a certain density (a
cubic foot weighs about one ounce and a-quarter)
dependent on the weight and pressure of the superincumbent mass ; but on a mountain top 15,000
feet high, as half the mass of the atmosphere is
below that level, (see " Pneumatics ") the air is
bearing but half the pressure, and consequently
has twice the volume of an equal quantity of air
at the sea-side, and a temperature consequently
many degrees inferior ; and the air which is at any
time on the mountain-top, may have been recently
before on an adjoining plain or shore, and in
gradually climbing the mountain side as a wind,
it must have been gradually expanding and cooling in proportion to the diminishing pressure.   It
is found that air, at first rising from the sea-shore,
becomes one degree colder for about 200 feet
of perpendicular ascent, and altogether about 50°
colder in rising 15,000 feet ; so that at this latter
elevation, water is frozen even near the equator,
where the temperature of low plains is at least
80°.   It thus appears that if a man could travel
with the wind so as to remain always surrounded
by the same air, he might begin his journey with
it from the summer vineyards of the Rhine, might
soon after find it the piercing blast of the alpine
summits ; and again, a little after, without any
change having occurred in the absolute quantity

of its heat, might feel it as the warm breath of
the flowers on the plains of Italy.

The explanation is here ready of why very elevated mountains in all parts of the earth are
hooded in perpetual snows. We have just said that
even at the equator, where the average temperature near the sea is 84°, water will be frozen
when carried to an elevation of 15,000 feet. A
line, therefore, traced on a mountain at this level
would divide the portion of it destined to sleep
under lasting ice and snow from the portion below
covered with green herbage. This line, wherever
found, is called the *snow line*, or *line of perpetual
congelation*. At the equator it is high in the atmosphere, because there is a difference of about
50° between the average temperature of the
country and the freezing point of water, *viz.* the
difference between 84° and 32°, and an elevation
of 15,000 feet corresponds to this difference; but
in a progress towards the poles, the line is met
with gradually nearer to the earth, as the difference in question is less. In Switzerland it is at
6,500 feet above the sea; in Norway, it is below
5,000. With respect to the line of congelation, it
is further to be remarked, that in tropical countries, because the temperature of the air is nearly
uniform during the whole year, the line or limit
of frost and snow is distinct and unvarying, that
is to say is narrow, particularly where the accli-
vity is considerable; but in countries to the north
and south, which experience strong contrast of
summer and winter, the line becomes broad and
less evident; because in the hot season much snow

is melted or half melted above what may be called an average line, while in winter much snow and ice are accumulated below this, to be melted again when summer returns.

In the breadth of the line of congelation for changeable climates, we have the reason of the formation of what are called *glaciers* around snow-capped mountains situated in such climates, and around such only. The snow near the upper part of the broad line having been only softened or half thawed in the preceding summer, becomes in winter, almost as solid as ice, and in the succeeding summer vast masses of this, detached by the action of the sun and of the central heat of the earth spreading outwards, and loaded with more recent accumulation of snow, are constantly falling down into the vallies beneath; where, being accumulated, and the crevices filled up with snow or with water which hardens to ice, they form at last the huge glaciers or seas of ice—*mers de glace*, which render certain regions so remarkable. The falling of such masses (called in Switzerland *avalanches*), is what renders the ascent to snow-clad mountains terrific and dangerous. Around Mont Blanc, in the awful solitudes of the elevated vallies, the avalanches are thundering down almost without interruption during the whole summer—in which season only the attempt to ascend the mountain can be made, and a pistol-shot, or any considerable agitation of the air, may suffice to set loose masses that will sweep away a whole convoy. Beneath glaciers there is always going on a melting of that part of the ice which is in contact with the earth, and hence a stream of water constantly issues from the bed of every glacier. These streams in Switzerland are the beginnings of the magnificent rivers the Rhine and Rhone. Like the avalanches breaking loose in summer among the mountains, there are in the polar seas vast masses of ice detached from the shores, and which afterwards move into warmer seas to be melted. These often become to the arctic bear rafts, on which, to his surprise, he finds himself voyaging into new latitudes, to be left at last adrift in the wide ocean when his ship has vanished from beneath him.

Although the proofs are not so immediately apparent, the line of congelation exists as truly every where in the open sky, over sea and plains, as where there are mountain heights to wear its livery ; and considerably below the line, the cold, aided by electrical agency, is sufficient to produce in the form of mist or clouds, a deposition from the air of the watery vapour contained in it. There is thus in nature an admirable provision to shade the earth at proper times from the too powerful rays of the sun, or to supply rain as wanted, without the transparency of the inferior regions of the atmosphere being much affected. As the watery vapour rising from sea or lake, and invisibly diffused in the atmosphere, can only reach to the height where the cold is great enough to condense it, the clouds may in general be regarded as the top of that atmosphere of watery vapour or aeriform water, which is always mixed more or less with the atmosphere of mere air; and as the quantity of watery vapour which can exist invisibly in a given space, depends altogether on the intensity of heat present, the clouds in a humid atmosphere will be low, and in a dry atmosphere will be high, or there may be none. An aeronaut mounting in his balloon through a clear sky, often approaches a dense cloudy stratum to plunge into it, and for a time to be surrounded with gloom almost of night; the face of earth being hidden from him below, while the heavenly bodies are equally veiled from him above; but rising still higher, he again emerges to brightness, and looks down upon the fleecy ocean rolled in mountain heaps beneath him, as the climber to a lofty peak may look down from the ever pure atmosphere around it on the inferior region of clouds and storms.

The diminished temperature of air in the higher regions of the atmosphere, often enables the natives of temperate climates, when led by circumstances to reside in tropical countries, where their health may suffer from the heat, to find near at hand, on some mountain height, the congenial temperature of their early homes. The author once, during a visit to the recently inhabited island of Penang in the strait of Malacca, examined this fact with pleasure not readily forgotten. The centre of the island is occupied by a lofty mountain ridge thickly wooded, on the northern summit of which a few residences, visible from the sea-shore like eagles' nests on a cliff, had just been constructed. Towards these, one morning at sun-rise, on an active little horse

of the country, and along a tolerable road, he
began to climb from the hot plain below. At
first there were around him purely tropical objects, inspiring tropical feelings,—the latter, modified indeed by the reflection that his track lay
through a forest, into which until lately the foot
of man had never ventured, and where the trees,
nursed through ages to their greatest growth, and
the stupendous precipices and the sublime water-
fall, &c. had so recently been exposed to human
observation ;—but as he gradually ascended, the
character of the vegetation was perceived to be
changing, and the air was becoming so light and
cool as irresistibly to awaken in him thoughts of
distant England—nay, almost the illusion of his
being there. At last, however, the summit being
reached, where a clear space opened to view
the whole country around, the attention was
quickly recalled to the fervid land of the sun.
The elevation is so great that at first the eye takes
account of only the grander features of the scene,
and such nearly as might be met with on a Grecian or Italian shore :—the expanse of sunny
water in that beautiful strait, stretching so far
north and south, the opposite continental shore
with its river winding seaward across the plain,
the town and the roadstead near it crowded with
ships, which appeared only as specks in a wide-
spread map, &c. ; but, on closer inspection, and
particularly with the aid of the telescope, were
descried the rich groves of cocoa-nut and banana,
the plantations of spice and cotton and sugar-
cane, the tawny labourers, the bamboo dwellings,

the fanciful canoes or prows, in a word, every
object bespeaking the torrid zone. Such then is
the scene, which even under the equator, an in-
valid by climbing a hill may place under his eye,
and where the thermometer near him stands as in
an English month of May.

The interiors of the islands of Jamaica and
Hayti have many situations of great extent, which
combine, as above described, the advantages of
tropical situation and temperate climate, and
English labouring colonists might well inhabit
the former. The vast plain of Mexico, and much
of the central land of South America, is similarly
circumstanced; and it is not uncommon, where
the ascent to the gigantic Andes is gradual, to
find at the bottom of the ridge a town, whose
markets are stored only with the productions of
the equator, while in a town higher up will be
seen only what belongs to the temperate skies of
Europe ;—climates of the earth naturally distant,
thus having met as it were in amicable vicinity
on the same rising plain.

Second. It might be anticipated that a dense
body, or one in which the constituent particles
may be supposed to fill more completely the space
occupied by it than the particles do in a rarer
body, would have smaller capacity for heat, in
proportion to the smaller space left vacant in its
mass : and in a general comparison of the capacities of *equal bulks* of different substances, such
anticipation is partly verified,—as when a pint of
dense mercury is found to have only about half
the capacity which a pint of lighter water has.

The accordance however is by no means universal, nor at all in proportion to the differences of density. Water, which is denser than oil, and according to the hypothesis should have less capacity, yet has in the same volume nearly double the capacity; and mercury, which being nearly fourteen times denser than water, might be expected to have only a fourteenth of the capacity, has really for equal volumes a-half, or, as formerly stated, for equal weights, a thirtieth.

Third. We are at last therefore compelled to admit, that the relation between various substances and heat, which we call capacity for heat, depends much more on the nature of the ultimate atoms of the substances than either on the absolute bulk or comparative density of the masses. Throwing much light on this subject, it has been ascertained in late times, that all material substances are composed of extremely minute unchangeable atoms, and of which, in different substances the comparative weights have been determined, although not the absolute weights; that is to say, for instance, the atom of gold is known to weigh four times as much as the atom of iron, although we do not know how many thousands or millions of atoms are required to form a grain of either. Now very recent researches seem to prove that for each ultimate atom, no matter of what substance, nearly the same quantity of heat is required to produce in a mass of the atoms a given change of temperature. Thus an ounce of iron which has four times as many atoms as an ounce of gold, has four times the capacity for heat.   The law seems to hold for all simple substances; and for compounds of these there seems to be another law not yet well made out.

Instead of the term *capacity for heat* used in the preceding pages, with respect to particular substances, that of *specific heat* has by some authors been preferred; but as the latter gives to a commencing student the idea rather of specific *kinds* of heat than of specific *quantities*, the term capacity has been here retained.

" *Each substance in nature, for a given change of temperature, undergoes expansion in a degree proper to itself, the expansion generally increasing more rapidly than the temperature, as the cohesion of the particles becomes weaker from increased distance, being remarkably greater therefore in liquids than in solids, and in airs than in liquids; the rate being quickened, moreover, near the points of change.*   (See the Analysis, page 6.)

The following table, containing the names of some common substances, solid, liquid, and aeriform, shews, by the figures following each name, how much the substance increases in bulk, by having its temperature raised from that of freezing to that of boiling water.   A lump of glass, for instance, would gain in the proportion of one cubic inch for every 416 cubic inches contained in it; while a mass of water would gain one inch or part in twenty-three; dilating thus for the same range of temperature nine times more than the glass.

Solids.

Glass gains one part in......... 416
Deal 416
Platinum 389
Steel 283
Cast iron........................... 278
Iron 271
Gold 221
Copper 194
Brass 177
Silver 175
Tin................................. 170
Lead 117

Liquids.

Mercury gains one part in...... 55
Water............................. 23
Oil of turpentine 14
Fixed oils 12
Alcohol 9

Airs.

Common air, ⎫
All gases   ⎬gain one part in... 3
and vapours ⎭

We have to warn readers here not to confound
the increase by heat of the general *bulk* of a solid
body with the increase of its *length*. The latter
is only one-third as great as the former. This will
be understood by considering that the increase of
bulk is divided between the *breadth* and *depth* (or
thickness) in common with the *length*. If the
substance of a metallic square rod or wire, for
instance, be dilated by heat, the hundredth part
of its bulk, it does not gain all that hundredth at

its end, becoming perhaps 101 inches long instead
of 100; but every part becoming deeper and
broader in the same proportion as it becomes
thicker (we may suppose it divided into a row of
equal little cubes), the rod gains in length only
the third part of an inch. A fluid enclosed in a
tube unchangeable by heat (if such tube there
were) would shew its whole dilatâtion in an increase of length, because there could be no swelling laterally, and its extremity would therefore
have a triple extent of motion from any variation
of temperature. A degree of this consequence is
obtained in our common thermometers, because
the containing glass, although dilatable by heat,
is so much less dilatable than the fluid within.
As regards solids, we have to inquire so much
more frequently respecting the dilatation in
length, breadth, &c.; that is to say, the *linear
dilatation* in some direction, than the increase of
general bulk, that tables are frequently made
stating only the linear dilatation. It may be
found at once from the above table, by recollecting that it is one-third of the increase of bulk :—
thus, if glass in passing from the freezing to boiling heat of water, dilate one part in 416 of its
bulk, it will dilate only the third of a part in
length, or a whole part in an extent of three times
416 or 1,248.

The expansion by heat of solids has been ascertained by bringing microscopic instruments to
bear on rods of the various substances heated to
various degrees, in troughs of oil or water. The
expansion of fluids, again, is found by filling a

glass vessel with a known weight of any fluid, and then ascertaining how much is made to run over or escape by a given increase of heat. This quantity, added to what is required to fill the increased dimensions of the heated glass vessel (which from the ascertained expansion of glass is known) forms the whole of the increase. It might be ascertained also by putting different liquids successively into the same thermometer-tube, and marking their comparative dilatations from changes of temperature examined by another thermometer.

The general and comparative expansion of solids by heat are exemplified in the following cases:

A cannon-ball, when heated, cannot be made to enter an opening, through which when cold it passes readily.

A glass stopper sticking fast in the neck of a bottle often may be released by surrounding the neck with a cloth taken out of warm water or by immersing the bottle in the water up to the neck: the binding ring is thus heated and expanded sooner than the stopper, and so becomes slack or loose upon it.

Pipes for conveying hot water, steam, hot air, &c., if of considerable length, must have joinings that allow a degree of shortening and lengthening, otherwise a change of temperature may destroy them. An incompetent person undertook to warm a large manufactory by steam from one boiler. He laid a rigid main pipe along a passage, and opened lateral branches through holes into the several apartments, but on his first admitting the steam, the expansion of the main pipe tore it away from all its branches.

In an iron railing, a gate which during a cold day may be loose and easily shut or opened, in a warm day may stick, owing to there being greater expansion of it and of the neighbouring railing, than of the earth on which they are placed. Thus also the centre of the arch of an iron bridge is higher in warm than in cold weather; while, on the contrary, in a suspension or chain bridge, the centre is lowered.

The iron pillars now so much used to support the front walls of houses of which the ground stories serve as shops with spacious windows, in warm weather really lift up the wall which rests upon them, and in cold weather allow it again to sink or subside—in a degree considerably greater than if the wall were brick from top to bottom.

In some situations (as lately was seen in the beautiful steeple of Bow-church, in London), where the stones of a building are held together by clamps or bars of iron with their ends bent into them, the expansion in summer of these clamps will force the stones apart sufficiently for dust or sandy particles to lodge between them: and then, on the return of winter, the stones not being at liberty to close as before, will cause the ends of the shortened clamps to be drawn out, and the effect increasing with each revolving year, the structure will at last be loosened and may fall.

The pitch of a piano-forte or harp is lowered in a warm day or in a warm room, owing to the expansion of the strings being greater than of the wooden frame-work ; and in cold the reverse will happen. A harp or piano, which is well tuned in a morning drawing-room, cannot be perfectly in tune when the crowded evening party has heated the room.

Bell-wires too slack in summer, may be of the proper length in winter.

One admirable contrivance for keeping the pendulum of a clock always of the same length, by making the greater expansion by heat of a middle bar of brass counteract the smaller expansion of two side rods of steel, was explained in vol. i., under the head of ' *Pendulum*,' as was also the construction of a balance-wheel having a corresponding property. A difference of a 100th of an inch in the length of a common pendulum causes a clock to err ten seconds in twenty-four hours, and a rise or fall of 25° of Fahrenheit's thermometer produces this difference. Another kind of compensation pendulum, distinguished by the name of its inventor *Graham*, is obtained by substituting for the solid bob or ball at the bottom a glass vessel containing mercury. The mercury on expanding by heat has its centre of gravity raised just enough to compensate for the lengthening of the rod of the pendulum.

Crystals do not expand quite equally in breadth and in length, and the expansion of one part may even cause a contraction of a part not yet warmed. The same is true of fibrous substances, as wood, which expands and contracts more in breadth than in length. This is proved by the leaking in cold weather of a ship's deck, which in warm weather is tight : an occurrence which the author once, in rounding the Cape of Good Hope, had to regret as the cause of destruction to some valuable specimens of natural history which he had collected among the eastern islands.

Other interesting examples of expansion and contraction in solids might be mentioned, but the above, in addition to what were given in vol. i. under the head of ' Repulsion,' may suffice. Bodies expanded by heat, unless when their intimate composition is changed by it, regain exactly their former dimensions on being cooled.

As is seen in the preceding table, the expansion of liquids is much greater than of solids.

A cask quite filled with liquid in winter, must force its plug in summer, or must burst : and a vessel which has been filled to the lip with warm liquid, will not be full when the liquid has cooled. Hence some cunning dealers in liquids try to make their purchases in very cold weather, and their sales in warm weather.

There exists a most extraordinary exception, already mentioned, to the law of expansion by heat and contraction by cold, producing unspeakable benefits in nature; viz. in the case of water. Water contracts according to the law only down to the temperature 40°, while, from that to 32°, which is its freezing point, it again dilates. A very curious consequence of this peculiarity is exhibited in the wells of the glaciers of Switzerland and elsewhere, namely, that when once a pool or shallow well on the ice commences, it goes on quickly deepening itself, until it penetrates to the earth beneath. Supposing the surface of the water originally to have nearly the temperature of the melting ice, or 32°, but to be afterwards heated by the air and sun, instead of the water being thereby dilated or rendered specifically lighter, and detained at the surface, it becomes heavier the more nearly it is heated to 40°, and therefore sinks down to the bottom of the pit or well; but there, by dissolving some of the ice, and being consequently cooled, it is again rendered lighter, and rises to be heated as before, again to descend; and this circulation and digging cannot cease until the water has bored its way quite through.

Airs are expanded by heat still more than liquids.

The expansion of aeriform bodies by heat produces many important effects in nature. Some of these have already been considered in preceding parts of this work, as, the rising of heated air in the atmosphere, causing the winds all over the earth; the same in our fires and chimnies supporting combustion, and ventilating and purifying our houses; the same again from around animal bodies, removing the poisonous or contaminated air that issues from the lungs, and insuring a constant supply of fresh air for the support of life, &c.

It is remarkable, with respect to aeriform bodies, that they are all equally dilated by the same change of temperature, receiving an increase of about a third part of their bulk ($37\frac{1}{2}$ parts in 100) on being heated from the freezing to the boiling point of water, viz. 180°, and their bulk being therefore doubled from the same standard point

by about 500°. This general truth holds, not only
with respect to the more permanent airs or gases,
but also with respect to all steams or vapours in
the dry state, that is, when not in contact with
the liquid producing them. The probable reason of
this uniformity is, that cohesive attraction, which
varies so much in different solids, modifying the
effects of heat upon them, in aeriform fluids does
not exist at all.

The extent of this dilatation for airs is so much
greater than for liquids or solids, that it forces itself much more strikingly upon the common attention. Thus a bladder containing considerably
less than its fill of air, becomes tense immediately
on being held to the fire. The air in a balloon
just escaping from a cloud has been so suddenly
expanded by the direct rays of the sun, as to have
injured the texture of the balloon ; and probably
some of the fatal accidents among aeronauts have
thus arisen. Burning fuel conveyed into a vessel or
case which can be suddenly and strongly closed,
will produce an expansion of the air confined with
it capable of bursting any ordinary material—in
short, will produce an explosion.

Now, if not before, at any rate soon after steam
engines began to be used, and had so strikingly
shewn to what important purposes the force of an
expanding aeriform fluid might be applied, the
thought would naturally occur that the force of
common air dilating by heat might also be rendered useful. Accordingly a variety of air-expan-
sion engines have been proposed, but as yet no
one has been reduced to profitable practice. Had
the truth been generally known, which very re-

cent investigations have proved, that any given
quantity of heat, when used to dilate air, produces
about four times the quantity of expansive power
that it does when used to form steam, the attempts to bring such an application of heat under
control would probably have been much more
numerous, and possibly by this time in a degree
successful. The subject is so interesting that we
shall subjoin a few remarks upon it.

To produce a cubic foot of common steam,
from water originally cold, about 1,150 degrees
of heat are required, as will be explained a few
pages hence. The same quantity of heat would
double the volume of about five cubic feet of atmospheric air,—as is known from the comparative
capacities for heat of the two substances, and the
rate of dilatation of air when heated. Now the
value for work of the foot of steam passing from
the boiler into a working cylinder would be, to
press up the piston of the steam-engine through a
foot, as from *c d* to *a b,* with a force all the
way of 15 lb. per inch of the piston surface; while
the working value of the five feet of air in dilating to double bulk would be to lift the piston five

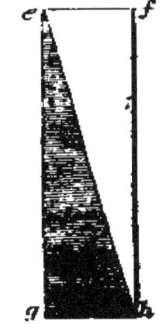

times as far as the steam, *viz.*
from *g h* to *e f,* but with a
force gradually diminishing
as the expansion went on,
from 15 lb. per inch at the beginning until the air had dilated to its destined volume,
when the force would altogether cease : its whole effect

:

therefore would be five feet impulsion of the piston, with a pressure the average between 15lbs.
and nothing, *viz.* 7½ lbs. per inch ;—and the friction in the two cases and the varying intensity of
the latter pressure being neglected, the force of
the air would be 2½ times as great as that of the
steam.   But it is farther to be considered, that
only about half the heat of a fire is applied to use
in the steam-engine, *viz.* that part which enters
the boiler, while the remainder passes up the
chimney ; and in an air-engine probably the
whole might be applied.   In an air-engine, moreover, there might be great increase of power from
the combustion, or semi-explosion of the inflammable gas evolved from the fuel.   As it was
easy, long before the steam-engine was contrived,
to determine the expansive force of steam, and to
compare it with any other force ; by proceeding in
a similar way with respect to heated air, we may
estimate its expansive power to be four times
greater than that of steam from an equal quantity
of fuel, and when used at a common or low pressure.   We see from this of what importance the
discovery would be of a means enabling us effectually to apply the force of expanding air.

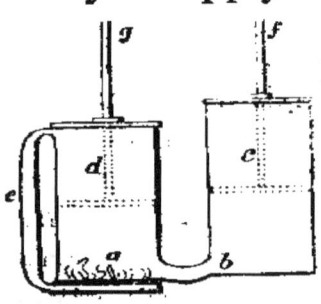

If we suppose a fire *a* to be
placed on a grate near the bottom
of a close cylinder, *d a,* and the
cylinder to be full of fresh air
recently admitted, and if we then
suppose the loose piston *g d* to
be pulled upwards, it is evident
that all the air in the cylinder

above *d* will be made to pass by the tube *e* through the fire, and will receive an increased elasticity tending to the expansion or increase of volume, which the fire is capable of giving it. If there were only the single close vessel *d a*, the expansion might be so strong as to burst it; but if another vessel *b c* of equal size were provided, communicating with the first through the passage *b*, and containing a close-fitting piston *c f*, like that of a steam-engine, the expansion of the air would act to lift the said piston, and by means of it might work water-pumps, or do any other service which a steam-engine can perform. At the end of the lifting stroke of the piston *f c*, it might be made to open an escape-valve for the hot air, placed in any convenient part of the apparatus, and to cause the descent of the blowing piston *d* to expel this, while a new supply of air would enter by another valve into the cylinder above *d*. The engine would then be ready to repeat its stroke as before, and the working would be continued as in a steam-engine.

The preceding simple conception of an air-engine occurred to the author's thoughts while considering the application of a condensed-air furnace to some chemical purposes. It appeared to him that the chief difficulties to be surmounted in applying any such engine to use would be, to prevent the very heated air and dust from injuring the valves and other working parts of the engine, and to obviate the inconvenience of the inequality of power at different parts of the stroke. Various expedients occurred to him. The overheating might be prevented by surrounding the cylinder, &c. with water; and both cylinder and piston would suffer less from dust if, instead of the common piston *c*, represented above, a great hollow plunger *a* were used (such as is here represented, and is now common in water-pumps for mines), embraced by an air-tight neck or collar at *b c*, which neck would be the only part of the cylinder requiring to be made with nicety. But a more complete security would be obtained by interposing water between the hot air and the piston, as represented in this other sketch, where the working cylinder *d*

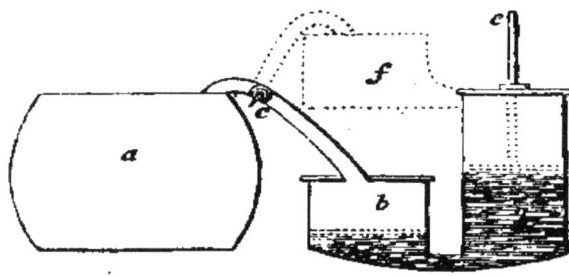

has a water-vessel *b* connected with it, and the heated air is admitted to press upon a float on the water-surface, to lift the working piston *d e*. This construction, too, if desired, would allow the fire-chamber *a* to be made larger than the cylinder, and to be kept constantly filled with highly expansive air, each discharge of which into the space *b* would be replaced by cold air either from the space above *d*, driven in through a tube as the piston ascended, or from a distinct blowing cylinder worked by the beam. And if it were wished to apply the same principle to an engine working with double strokes, that is, forcing the piston alternately up and down, as in the double stroke steam-engine, the object might be attained, by having a second water vessel *f,* communicating with the part of the working cylinder above the piston *d*; and the air would pass alternately to the one or the other vessel *b* or *f,* by the operation of the cock *c,* as steam passes in a steam-engine : the supply of fresh air to the chamber *a* would be given by a blowing cylinder worked through a connexion with the engine, as the air-pump of a steam engine is worked.

The sketch of an air-engine, as here given, was included in the specification of a patent for another object engaged in some years ago by a friend of the author's ; but he being almost immediately called to other business, and the author's professional engagements precluding his attention to the subject, it was not prosecuted. In the specification, drawn up by an engineer in town, some minor adaptations were described One experiment has lately been made by a Swedish engineer with the simple form of dry apparatus described at page 72, for the purpose of ascertaining its power, and the effect was found to be several fold greater than of steam from the same quantity of fuel ; but the apparatus was rude, and only calculated to prove in a short trial, the existence of the power, but not the fitness of the machine to endure long uninjured, and to be rendered easily obedient to control : a complete experiment therefore remains still to be made. Could an obedient and durable engine be contrived, at all approaching in simplicity to the plan given above, its advantages over the steam-engine would be very considerable. First, its original cost would be much less, by reason of its small comparative size, its simplicity, and the little nicety of workmanship required.   Secondly, it would occupy much less room, and would be very light ; hence its peculiar fitness for purposes of propelling ships and wheel carriages.   Thirdly, the quantity of fuel required being so much less, would not load the ship or carriage, leaving little room, as happens in steam-boats, for any thing else.   Fourthly, the expense of fuel and repairing would be little. Fifthly, the engine could be set to work in a few minutes, where a steam-engine might require hours.   Sixthly, little or no water would be required for it.

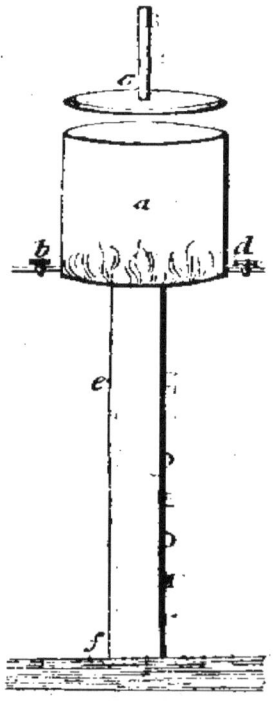

Another modification of air-engine, called a *gas vacuum engine*, has lately been proposed, and many expensive trials have been made of it ; but it is in its nature a most wasteful machine, evidently throwing away at least nine-tenths of the power which its principle generates. It was of this nature in an experiment which the author witnessed. A little of the common coal-gas was admitted by the cock *b* at the bottom of the cylinder *a*, and was there inflamed, the lid *c* being at the time raised. The combustion rarified the lower stratum of air, so that the air above was expelled, and about one-fifth of the original contents of the cylinder was made to occupy the whole. The lid was shut down as nearly as could be judged at the moment of greatest expansion, so that when the small portion of air and vapour remaining was again cooled, the interior of the cylinder approached nearly to the state of vacuum. It, in fact, retained only a fifth of the air. A communication being then opened by the tube *e* from the vacuous cylinder to a water reservoir ten feet below, the water was driven up by the atmospheric pressure, and filled more than half of the cylinder. The water so raised was then made to turn a common water-wheel, and so to do work.—A larger quantity of water, however, could be raised to the same height at less expense by a steam-engine. The proposer also hoped that he would be able to make the atmosphere pressing into his imperfect vacuum, act directly upon a piston as steam does, and with power cheaper than that of steam; but in this anticipation too he was completely in error. To produce his imperfect vacuum cost him very nearly at the same rate as it costs to produce the perfect vacuum in a steam-engine, and his vacuum for equal bulks was worth, as a working power, only about one-fourth as much as the steam vacuum. This may be understood by considering that in a perfect vacuum a piston rises all the way with the same force, which if common steam be used, is 15 lbs. per inch (the piston may be supposed to rise

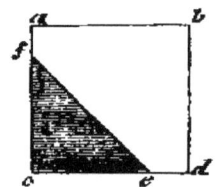

from *c d* to *a b*), but if the vacuum were only three-fourths towards being perfect, the pressure on the piston would be only three-fourths of 15 lbs. at the commencement of the stroke, would then rapidly diminish, and would have ceased altogether when the piston had made three-quarters of its journey, or to *f.* The force in the first case would be represented by the whole line *c d* and the square space *c d b a*, and in the second by the shortening lines and the triangular space *c e f.*

On considering the foregoing diagrams we may perceive that in the vacuum-engine, by far the greater part of the force produced by the combustion of the gas is absolutely wasted, or put to no use, *viz.* the whole expansive force during the sudden combustion or explosion. It is evident that if a tenth part of the aeriform contents of a cylinder acquire elasticity enough,—and a fourteenth part in a nice experiment does so—to be able afterwards to occupy the whole cylinder, it must begin its expansion with the force of a ten-fold atmospheric condensation, that is, a pressure of 150 lbs. on the square inch of a piston withstanding it, which pressure will then gradually diminish as the piston rises, but will amount to an average of five times the atmospheric pressure, or 75 lbs. per inch all the way ; being therefore quadruple or more, that of steam against a perfect vacuum, and therefore again, by our former calculation, at least twelve times greater than the

force obtained from the imperfect vacuum of the
engine under consideration.

It is a question which the author thinks will one
day be answered in the affirmative, whether nearly
the whole force of exploding gas may not be converted into a calmly working power, producing
from a given expenditure, ten times or more the
effect obtained in the vacuum engine described
above, and therefore more than from a steam-engine
incurring the same expense.   There are probably
various ways in which the object may be attained.
The following sketch is offered merely to give the
reader an idea of a machine for such a purpose.

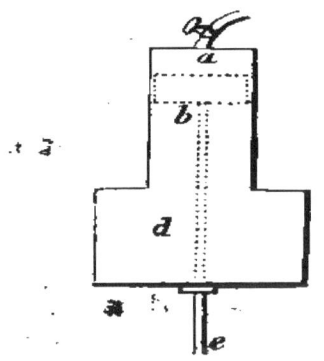

Suppose *b* to be a very heavy
close-fitting piston sliding in
the cylinder around it, and
suppose the space *d* open to
the cylinder, to be filled with
atmospheric air of double or
greater density ; then if a
mixture of explosive gases admitted by a cock to the chamber *a* (formed between the
piston and end of the cylinder) be inflamed, the
heavy piston will be shot forward, like a cannon ball, against the condensed air in *d* ; and
owing to the momentum acquired in the first instance, it will advance much beyond the point
where the exploded gas and air in *d* would balance each other at rest:—the quantity of gases
admitted would be just such as to carry it to the
end of the cylinder.   The piston rod *e* would
then by a catch or ratchet, be connected with the

work to be done,. and after the condensation of the exploded gases in the cylinder, would be pressed back again, with the double or greater than atmospheric force in *d*, as if urged by high pressure steam. The first figure at page 74 represents a form of cylinder which might also answer for this purpose, the heavy plunger being thrown up, to work by its weight in descending.

It is to be remarked that the first modification of air engine described at page 72, is partly an explosive engine such as contemplated above, for the gas separated from the coal during the moment of slackened combustion while the lately used air is passing out, becomes an explosive accumulation for the fresh air about to enter. The trial alluded to above proved this to be the fact.

" *The expansion of bodies by heat increases more rapidly than the temperature, and particularly near the melting and boiling points, that is, their points of changing into liquid or air, being however exactly proportioned to the temperature after the change into air.*" (See Analysis, page 6.)

If a given quantity of heat, of that for instance contained in some measure of boiling water or of common steam, be added to a mass of cool water, it will produce in this a certain increment of bulk ; and if other equal quantities of heat be afterwards successively added, under the nice management which such an experiment requires, each new addition will produce a greater increment of bulk than the preceding, particularly when the water approaches to boiling ; but after the water is converted into steam, any farther increase of bulk will be exactly proportioned to the increase of temperature. The same truths may be proved by the converse experiment of abstracting successively equal quantities of heat from steam or water (as by making it melt equal quantities of ice), and noting the rate of contraction. What is thus true of water in relation to heat is true also of bodies generally, each however having a rate of expansion and temperatures for melting and boiling proper to itself. The quickened rate of expansion in solids and liquids might have been anticipated from reflecting, that each successive quantity of heat added to a mass, meets with less resistance to its expanding power than the preceding, owing to the diminishing force of cohesion of the particles as the mass expands : while in an air or gas, again, as cohesion has altogether ceased, each addition of heat is at liberty to produce its full and equal effect. If the capacity of substances for heat did not increase with their bulk, the terms " increase of heat " and " increase of temperature " would have the same meaning, and this subject would be more simple.

The reflection will naturally occur here, that as in the common thermometer the mercury must rise or expand more for a given quantity of heat added at a high temperature than at a low temperature, the scale should be divided to correspond with the inequality. Now this reasoning is good, but the difficulty of complying with it in practice is such, that the inconvenience of the

G

slight error arising from an equal division is submitted to. An air thermometer with equal divisions is very correct, but from wanting many of the advantages of the mercurial thermometer is little employed; and fortunately it happens that in the mercurial thermometer there is such a counterbalancing relation between the expansion of the mercury and of the containing glass, as to render the error alluded to, at least for any middle range of temperature, very trifling. The subject of unequal thermometric dilatation in the same liquid, and of differences in different liquids, depending on the proximity to their boiling points, &c., is well illustrated by Du Luc's experiment of filling different thermometer-glasses with different liquids, and noting their comparative indications when heated through the same range of temperature. He marked on each the points at which the liquid stood when the glass was placed in freezing and in boiling water, and then divided the intervening spaces into eighty parts or degrees. The discordance of their dilatations is here detailed.

Mercury.	Olive Oil.	Alcohol.	Water.
0	0	0	0
10	9.5	7,9	0.2
20	19.3	16.5	4.1
30	29.3	25,6	11.2
40	39.2	35.1	20.5
50	49.2	45.3	32
60	59.3	56.2	45.8
70	69.4	67.8	62
80	80	80	80

The singular discrepancy in the case of water is owing to the peculiarity described in former pages, of its contracting by cold only down to 40° of Fahrenheit, and then dilating again until it freezes.

Laborious investigations have been made by the French chemists to discover a comprehensive law determining the rate of expansion in all bodies, but the object is not yet satisfactorily accomplished.

" *To melt a solid body, or to vaporize a liquid, a large quantity of heat enters it, but in the new arrangement of the particles and generally increased volume of the mass, the heat becomes hidden from the thermometer and is called* LATENT HEAT. *It re-appears during the contrary changes, after whatever interval.*" (See the Analysis, page 6.)

THE expansion of bodies by heat, instead of proceeding throughout in some nearly uniform or gradual manner, exhibits in its course two singular transformations of the body, *viz.* when the solid breaks down into a liquid, and when the liquid swells out into an air or gas. The substance of water, for instance, when at a low temperature, exists in the solid form called *ice ;* but at 32° of Fahrenheit it becomes liquid or *water,* and at 212°, even under the resisting pressure of the atmosphere, it suddenly acquires a bulk nearly 2,000 times greater than it had as a liquid, being then called *steam* or aeriform water. And other bodies under analogous circumstances undergo similar changes. It is further remarkable, that although during the changes a large quantity of heat enters the mass, producing in the one case liquidity, in the other the form of air, the temperature is the very same, immediately after, as it was before. Water running from melting ice affects the thermometer but as the ice does, and steam over boiling water appears no hotter than the water. The glory of originally making the discovery of the facts now referred to by the terms *latent heat*, or *caloric of fluidity*, belongs to the illustrious Dr. Black. The construction of the modern steam-engine was an early result of kindred investigations made by Dr. Black's friend, James Watt.

We select the following instances as serving to display the subject of *latent heat* in its various bearings.

A mass of ice brought into a warm room, and therefore receiving heat from every object around it, will soon reach the temperature of melting or 32°, but afterwards both the ice and the water formed from it will continue at that temperature until all be melted :—the heat which still continues to enter, effecting a change only in the form of the mass. And in the case supposed, whatever time was required for heating the mass of ice *one degree*, just one hundred and forty times as much will be required for melting it ; proving that 140° is the latent heat of water.

If two similar flasks, one filled with ice at 32°, and the other with water at 32°, be placed in the same oven or over the same flame, the water will gain 140 degrees of heat while the ice is merely dissolving into water at 32°: and in the course of the experiment, a correspondence will always exist between the phenomena; for instance, when the water has gained 14° of heat, it will be found that just a tenth part of the ice is melted.

If equal quantities of hot and cold water be mixed together, the whole acquires a middle temperature, each degree lost by the hot water becoming a degree gained by the cold: but if a pound of ice at 32°, and a pound of water 140° hotter be mixed together, the 140° of heat will go merely to melt the ice, for there will result two pounds of water at 32°.

If a flask of water at 32°, or its freezing point, and a similar flask of strong brine at 32°, but which does not freeze until cooled to near zero, be exposed together in the same cold place, it will be found that when the brine has lost 10° of its heat the water flask will still exhibit an undiminished temperature, but a fourteenth part of its contents will be converted into ice. Now as in such a case the water flask must continue to radiate away heat just as much as the other, it can maintain its temperature only by absorbing into its general mass the heat which was latent in the portion of water frozen.

It is possible by cooling water slowly and in perfect repose, to lower its temperature, while yet liquid, ten degrees below its ordinary freezing point; but then, on the slightest agitation, ice will be formed. It might be expected in such a case, that the whole water would instantly freeze, because all colder than common ice; but no, only a fourteenth part does so, and singularly, both that fourteenth and the remaining liquid are rendered in the moment ten degrees warmer. Here the 140° of latent heat escaping from the fourteenth part which freezes, become 10° of sensible heat for the whole mass, so that the remaining water has the temperature at which it only begins to freeze.

Strong solutions in hot water of various neutral salts, if allowed to cool while exposed to atmospheric pressure, soon deposit crystals of the salts; but in a close vessel protecting them from such pressure, they will remain liquid even when cold. Now at the moment of opening such a bottle, the salt immediately crystallizes, and the latent heat given out by the solidifying particles warms very sensibly the remaining liquid and the bottle.

From the preceding facts it may be perceived, that the quantity of ice formed or melted in any case, becomes a correct measure of the quantity of heat transferred. From this consideration, the illustrious Lavoisier constructed his calorimeter, or heat-measure. It is a case or vessel lined with ice, and the quantity of heat given out by any body placed in it is indicated by the quantity of water collected from the melted ice.

Had the latent heat of water been only 1° or 2° instead of 140°, the earth, except in its tropical regions, would have been scarcely habitable. The cold of a single night might have frozen an ocean, and the heat of a single day might have converted the accumulated snows of a winter into one sudden and frightful inundation. As the fact is, however, both changes are beautifully gradual, and easily controlled or prepared for.

The fact of latent heat in other liquids than water is familiarly exhibited in the slow melting —of the metals; lead or pig-iron for instance—of butter or oils—of glass, &c.; and on the other hand, in the slow solidification of the melted masses when heat is again abstracted.

The substances below enumerated, while passing from the solid to the liquid state, absorb and render latent the quantities of heat here noted:

Ice	140°
Mercury	142
Bees'-wax	170
Tin	442
Zinc	492

If a piece of frozen mercury (the temperature of which is at least 40° below zero) be thrown into a little water, the latent heat of the water immediately passes into the mercury and melts it, but then the water is so cooled as to become ice.

" Latent heat of *aeriform fluids.*"

Water in a vessel placed over a fire gradually attains the boiling temperature or 212°, but its temperature then rises no more, because the farther addition of heat becomes *latent* in the steam escaping during the ebullition. The quantity of heat which becomes latent in steam is discovered by noting how much more time is required for boiling a quantity of water to dryness, than for merely heating it a certain number of degrees. The experiment indicates about 1,000° ; that is to say, that 1,000 times as much heat is latent in any quantity of water formed into steam as would raise the temperature of the liquid water one degree. Watt had found that water in a vessel placed over a lamp was about six times as long in being completely evaporated, as in being originally heated from an ordinary temperature to that of boiling.

If we place in the same oven or over similar flames two like vessels containing water, one of which vessels is open at top and the other is strongly closed, the two will gain heat equally up to the boiling point, but afterwards the open vessel from giving out steam will remain at the same temperature, while the other, by confining all the heat that enters, will shew the temperature rising as before, until the increasing tendency of the water to dilate forces the vessel open. Supposing the water in the latter vessel, before vent is given, to have become 100° hotter than common boiling water, instead of the whole being immediately converted into steam as might be expected, only a tenth part will be so changed (the same quantity as will be found to have already escaped from the other vessel), for the tenth part requiring in the form of steam 1,000° of latent heat, will take the excess of 100° from the other nine parts, and will leave them as common boiling water. If water heated considerably beyond the boiling point be allowed to expand very suddenly, the whole is blown out of the vessel as a mist, by the steam formed at the same instant through every part of the mass, but the whole mass in such a case is no more converted into true steam, than the whole of very brisk *soda water* is converted into air when similarly thrown out by the sudden extrication of the carbonic acid gas, on uncorking the bottle. Misconception of this matter has led to most wasteful experiments on steam-engines of very high pressure.

The same indication of the latent heat of steam is obtained by the converse experiment of first converting a quantity of water into steam, and then admitting it to cold water or to ice. A pound of steam will raise the temperature of ten pounds of cold water 100 degrees, or will melt about 8⅓ pounds of ice.

In the great quantity of heat which becomes latent in steam, we perceive the reason why water projected upon a raging fire so powerfully represses it :—and hence again why *fire* and *water* are so often adduced proverbially as furnishing a striking contrast.

It was when Watt had discovered how much heat was lost by losing steam, that he contrived the separate condenser for his steam-engine, by which he saved three-fourths of the fuel formerly used.

Substances differ among themselves in regard to the latent heat of their vapours as much as in their other relations to heat. Thus the latent heat of the vapour or steam of :

Water........ is ......1,000°
Vinegar	900
Alcohol	442
Ether	300
Oil of turpentine...	177

From the less latent heat in other vapours than in that of water, we might at first suppose that there would be great advantage from using them in steam-engines. Accordingly numerous experiments have been made, and patents secured under this idea; but the fact is, that in the same proportion as the heat is less, the volume of the vapour is less, and therefore no mechanical advantage is obtainable.

The influence of external pressure in keeping the particles of liquids together in opposition to the repulsion of heat seeking to render their mass aeriform, was considered in the chapter on ' *Pneumatics* ;' but to make the present section complete, the subject must be here shortly resumed.

Because any liquid, water for instance, while receiving heat remains tranquil, and apparently unchanged, until it reaches the boiling point, at which bubbling or conversion into vapour takes place, we might suppose its ordinary boiling temperature necessary to enable it, under any circumstances, to assume or to maintain the form of air. But this is no more true than that a common spring compressed against an obstacle has no tendency to expand or recover itself until the obstacle happen to give way. Liquid water with its heat is really a spring much compressed by the weight of the atmosphere, and seeking to expand itself into steam with force proportioned to its temperature. Even at 32°, or its freezing point, if placed in a vacuum, it assumes the form of air, unless restrained by a pressure of $1\frac{1}{2}$ ounce on each square inch of its surface ; and at any higher temperature the restraining force must be greater ; at 100°, for instance, it must be 13 ounces ; at 150°, 4 lbs. ; at 212°, 15 lbs. ; at 250°, 30 lbs., and so on :—and whenever the restraining force is much weaker than the expansive tendency, the formation of steam will take place so rapidly as to produce the bubbling and agitation called *boiling*. Now it is because the atmosphere or ocean of air which surrounds the earth happens to have in it 15 lbs. weight of air over every square inch of the earth's surface, and presses on all things there accordingly, that 212° is called the boiling point of water. An atmosphere less

heavy would have allowed liquids to burst into vapour at lower temperatures, and one more heavy would have had a contrary effect.—The exact degree of expansive force for every degree of temperature in water and other liquids, has been ascertained by heating them in vessels furnished either with properly loaded valves, as at f in this figure, or with a tall upright

tube, as *d b*, into which the liquid *c* may force
a column of mercury to an elevation marking
the expansive tendency ; the valve and mercury
being of course protected from the external atmospheric pressure, or the necessary allowance being
made for that pressure. Boiling at the bottom of
a deep vessel is resisted by the weight of the
liquid in addition to that of the atmosphere, as
already explained, and consequently the temperature at which it occurs is there higher than near
the surface of the vessel. Boiling heat is greater
also in other cases, as—in a deep mine, where of
course there is additional depth and weight of atmosphere over any exposed liquid,—at times when
the barometer is unusually high, that is to say,
when the atmosphere is unusually heavy—in
cases where air or steam is confined over the boiling surface so as to press more upon it, as when
brewers for a time shut the lid or valve of their
great boilers, &c. Water placed on the fire in a
strong vessel, from which steam cannot at all escape, may be rendered even red-hot, without a
bubble forming or one particle being dissipated ;
but the tendency to expand into steam is then
great enough to burst any known material of moderate thickness. The Marquis of Worcester exploded a cannon by shutting up water in it, and
then surrounding it with fire.—Boiling temperature is lower again when the experiment is made
on mountains or in other situations above the
level of the sea, where there is less height of air
resting over the boiler. In the city of Mexico,
which is 7,000 feet above the sea, water boils

before it reaches the heat of 200°, instead of, as in places near the sea-level, at 212°. Wollaston's thermometer, beautifully adapted for determining the height of mountains, balloon ascents, &c., by merely indicating the heat of boiling water in any situation, is a fine illustration of this truth. If in any place we take off the atmospheric pressure from a liquid, as by placing it in the receiver of an air-pump, it will boil at very low temperatures indeed. Water thus treated boils at 70°, which is 20° below the heat of some English summer days, and ether boils when colder than common ice. Generally, in a vacuum, substances boil at a temperature 124° lower than while restrained by the atmospheric pressure.

Consequences of these truths respecting the boiling temperature, are the following.

As water when heated dilates itself in the form of steam, the steam presses on a given extent of surface with the same force as the water itself would do; and in a steam-engine, the temperature of the water tells the degree of force which the steam is exerting on the piston.

Because in the case of steam, the same law holds as for aeriform fluids generally, *viz.* that the outward elasticity or spring increases in proportion as the fluid is more condensed—high-pressure steam is merely condensed steam, just as high-pressure air is condensed air; and to obtain a double or triple pressure, we must have twice or thrice the quantity of steam under the same volume.

The reason that high-pressure steam issuing

from a boiler heated perhaps to 300°, is not hotter than low-pressure steam from a boiler at 212°, is, that in the instant when the high-pressure or condensed steam escapes into the air, it expands until balanced by the pressure of the atmosphere, that is, until it become low-pressure steam, and it is cooled by the expansion, as air is cooled on escaping from any condensation.

The vessel called *Papin's Digester*, is merely a pot, which can be kept closed in spite of the force of the steam formed within it; and in such a vessel water can be heated far beyond the ordinary boiling point,—enough, for instance, to dissolve and extract all the gluten or jelly of bones, and to form from them a rich soup where common boiling would procure nothing :—or even enough to melt lead.

The person who urges a fire under a boiling pot with the hope of making the water hotter, is foolishly wasting the fuel, for the water can only boil, and it does so at 212° of the thermometer.

As different substances under a given pressure, become aëriform at different temperatures, mixtures of such may be decomposed by heat. If a mixture of spirit and water, for instance, be placed over a fire, the spirit will boil off long before the water. If the spirituous vapour be caught apart and condensed, it is said to have been *distilled*. Other distillations are of the same nature.

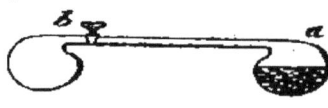

The instrument here represented consists of a glass tube blown into bulbs at the two ends *a* and *b*, and hermetically sealed, after receiving into it some water, but no air. If one of the bulbs be heated more than the other, the steam or vapour in that one will, for the reasons stated above, be denser and stronger than in the other, and will therefore force its way into the other; but, owing to the lower temperature there, a part of it will be condensed, making room for more. Hence, if the difference of temperature between the bulbs be long maintained, the whole water will by a sort of distillation gradually pass into the colder bulb. If the difference of temperature become at any time considerable, the liquid will boil in the warmer bulb, even although the source of heat be only the living hand grasping it.

To the author it appears that by a larger apparatus made on this principle, fresh water might be conveniently obtained from salt-water on board ship, or on an island having no fresh springs. Suppose any two air-tight vessels like a and b, communicating by a tube furnished with a stopcock near b, then if the vessel a were filled with salt-water, and were heated by being exposed to the sun (its surface being blackened and protected by glass from the cooling effect of the air), and if the other vessel b, after being also filled with water, were made a vacuum by pumping the water out from the bottom, and were then kept as cold as possible by wetted coverings and a current of air,—on opening the cock at b, vapour would pass over from the heated vessel to be constantly condensed in the colder, and there would be a distillation from the sea-water of perfectly sweet water by the natural action of the sun alone. Cases have occurred where a knowledge of this fact would have saved ship-wrecked crews from perishing by thirst; and there are rocky islands in the ocean which would become pleasantly habitable by the adoption of such a means, where now there is no supply of fresh water, but from precarious rain or importation from abroad.

When a substance has reached the temperature at which it boils, that is to say, at which its vapour becomes a balance to the atmospheric pressure, its dilating force is strong indeed. Persons may not reflect that 15 lbs. on a square inch is about a ton on a square foot,—and such is the power with which the vapour of all boiling substances rises from them—sufficient in a single Cornish steam-engine to urge the piston with the power of 600 horses! But the tendency to expand at temperatures much below boiling, is still, as already stated, very great, and although not attracting common attention, is silently working many beautiful and important ends in the economy of nature. As into a perfect vacuum, freezing water gives out a steam or vapour that would lift with force of 1 lb. on per inch, or 16 lbs. on a square foot, and even solid ice gives out its vapour of nearly equal strength; so also do other liquids and solids. There is an aeriform mercury, dense in proportion to the temperature, in the apparently empty space called the Torricellian vacuum, over the mercury in a barometer tube; and around camphor, all the essential or volatile oils, &c., there is similarly an atmosphere of the substance in the form of air.

It had for a considerable time been known that into a perfect vacuum bodies emitted almost instantly, in the form of air, a quantity of their substance proportioned to their temperature; but it was reserved for Mr. Dalton to make the admirable discovery, that even into any space filled with air these vapours arise in quantity and density the same as if air were not present—the two fluids seeming to be independent of each other, with the exception that in a vacuum the equal diffusion of a vapour takes place at once, while in a situation already occupied by air, it proceeds only as the vapour can force its way through the particles of the air, and in general takes place by a tranquil evaporation from the surface instead of the agitation of ebullition. In an apartment with an open vessel of water in it, there is soon, although invisible, a steam or watery vapour mingled with the air, as dense as if the room were a vacuum at the same temperature.

Consequences of this important truth are the following.

That it is only an atmosphere of the substance of each body, which by pressing on it can prevent its further dissipation by heat. Thus we can only save camphor, musk, smelling oils, spirits, water, &c., by placing them in closed bottles or vessels, in which, additionally to the air present, an atmosphere of their own substance is soon formed, in-

H

volving the remaining masses with pressure proportioned to their temperature and its density.

The important process of drying things is merely the placing them under an elevated temperature if attainable, and in an atmosphere not containing so much of the liquid as to be saturated at the temperature. The effect of wind or motion of the air in quickening evaporation, is owing to its removing air saturated with the moisture, and substituting air which is not—thus producing nearly the case of the substance placed in a vacuum.

If air at a certain temperature, contain mixed with it as much water as can be sustained in the form of invisible vapour at that temperature, and if by any cause, as by rising in the atmosphere, the air be then cooled, it will abstract heat from the vapour, and cause a portion of this to be precipitated or visibly condensed into a fog or rain. Water rising as invisible vapour from the surface of a lake or river, when it has reached a certain height, is condensed into the stratum of clouds, which at times usefully protect the fields from the intense meridian sun, and may fall again as refreshing showers over the country.

It is the tranquil and invisible evaporation of which we are now speaking, which lifts from the surface of the wide ocean all the water which, after condensation, is again returning to it in the myriads of river streams which give life and beauty to the face of nature.

In warm climates there are inlets of the sea, shut off occasionally from the parent ocean, and where, after the sun's rays have drank up all the water, the deposited salt remains to be carried away in loads for the uses of man, as sand is carried from any ordinary shore. There are in the bowels of the earth prodigious accumulations of salt, formed doubtless in the same way, during the revolutions of the antediluvian world, and now explored as salt-mines. When the Nile overflows its banks with waters, dissolving, although in almost imperceptible proportion, mineral substances brought from central Africa, and fills reservoirs afterwards dried up by the sun's heat, it leaves in these a rich store of crystallized natron or soda.

The following are other instances of vapour, invisible while at a higher temperature, being thickly precipitated when air, with which it is mixed, is cooled, or when it touches a colder solid body :—the steam observed at night and morning hovering over brooks and marshes heated by the sun during the day :—the frost-smoke, as it is called, which lies on the whole face of the Greenland seas in the beginning of winter, where the water warmed by the long day of the polar summer, continues to emit its vapour for a considerable time after summer is past, into an atmosphere become too cold to preserve it invisible :— the breath or perspiration of animals, of horses in particular after strong exertion, becoming so strikingly visible in cold and damp weather, or even in warm weather, when the air is already charged with moisture :—in cities where there are deep drains communicating with kitchens, manufactories, &c., and constantly filled with moist and warm air; the vapour-loaded air, although clear or transparent below, immediately on escaping into a frosty atmosphere, lets go its moisture, with the appearance of steam issuing from a great subterranean cauldron. Steam over water in any boiler is transparent or perfectly aeriform—as may be seen when water is made to boil in a vessel of glass, but as soon as it is cooled by contact or admixture of colder air it ceases to be true steam, and is condensed into small particles of water suspended in the air. Many persons, while thinking of steam, figure it only in this latter state, as particles of water mixed with air nearly as a subtle powder might be mixed, and its substance occupying really no more space than the original water did. Now until steam is cooled and condensed, it is of a nature to fill alone any appropriate vessel and powerfully distend it, just as air fills and distends a bladder. Steam issuing from the spout of a kettle is hardly seen near the mouth, but as its distance from the spout increases, it is cooled into a thick cloud or vapour.

In a vessel from which air and atmospheric pressure are excluded, even the temperature of freezing water being sufficient to maintain permanently in the state of gas or air, many substances which exist as liquids under the atmospheric pressure,—and the whole mass of such a substance when placed in a vacuum, not being instantly converted into gas because the portion which first rises becomes an atmosphere weighing upon the remaining mass, and because, moreover, that portion, by absorbing from the mass much heat into the latent state, cools the mass much below the freezing point;—we see why the liquids now spoken of are so rapidly cooled to at least the freezing point if placed where a vacuum can be maintained, that is to say, where, after common air has been removed, the aeriform matter rising from them, and absorbing their heat, is promptly and in a continued manner abstracted. It is thus that water placed in the exhausted receiver of an air-pump is so rapidly cooled, and that when there is beside it a vessel of concentrated sulphuric acid, or other substance capable of absorbing the watery vapour as formed, it is soon reduced to the state of ice; or again, that water, or even mercury, surrounded by ether evaporating in a vacuum, is so quickly frozen. It is thus also that if one bulb of the instrument described at page 99 be immersed in a freezing mixture, the water in the other and distant bulb will soon become ice; for the vapour rising from that water into the vacuum maintained throughout the apparatus by the freezing mixture, is immediately condensed again in the immersed bulb, and leaves the vacuum still free for the ascent of more vapour, to carry away more heat from the water as latent heat, and to make it freeze.

As we have explained also, that in a liquid there is the same tendency to evaporate whether it be or be not exposed to the air, we see the reason why all evaporation is a very cooling process. The effect however in air, is neither so rapid

nor so great as in a vacuum ; first, because the
presence of the air impedes the spreading from
the liquid surface of the newly formed vapour,
and keeps it where its pressure resists the formation of more vapour ; and, secondly, because the
air in contact with the liquid, shares its higher
temperature with the liquid.   Still in India flat
dishes of water, placed through the night on beds
of twigs and straw kept wet and in a current of
air, soon exhibit thin cakes of ice—and thus ice is
procured in India for purposes of luxury.

The absorption of latent heat in the evaporation
which goes on from the sea and earth in all warm
climates, greatly tempers the heat of these climates, and the vapour afterwards spreading to
the poles, as explained in ' *Pneumatics*,' under
the head of *winds*, carries warmth thither to be
given out when it is re-condensed into the form of
rain, or is solidified as snow.   The formation any
where of mist or rain warms the air most sensibly,
by the liberation of the latent heat from the precipitated vapour.   Again, the liquid water which
during winter is converted into snow or ice had
been a reservoir of latent heat stored to temper
the frosty air of the commencing cold season ;
and in the following spring, such ice and snow
serve as empty receptacles in which the first violence of the returning sun hides or expends itself ;
allowing the temperature to change more gradually, and for many living beings therefore more
safely.   The vast stores of ice and snow among
high mountains, as among the Alps and Pyrenees,
are often stores of mild temperature to regions

around ; for besides cooling the air near them, they are the never-failing sources of the rivers which run from them during the whole of summer, carrying freshness through the lands :—from the Alps, for instance, proceed the Rhine and Rhone—most romantic and beautiful of European streams ; and from the Pyrenees, the little Gâve, &c., which while channels around from lower regions are almost dried up by the summer heat, flows only the more freshly as the heat is greater, and the feeding snows are more abundantly dissolved.

Men in artificially raising temperatures are generally causing the liberation of heat which had been previously latent, and in lowering temperature or producing cold, they almost solely effect their purpose by rendering a quantity of heat latent.

Lavoisier thought that the heat of all combustion was merely the latent heat of the oxygen gas concerned in the combustion, given out during its combination with the burning body. It is so in part, but we now know that it depends more on the intensity of the chemical action between the combining substances. The water thrown upon quick lime to slake it, becomes solid in combination with the lime, and gives out its latent heat so remarkably as often to set fire to a wooden vessel or ship containing it.

When dwelling-houses, green-houses, manufactories, &c. are warmed, as is now common, by the admission of steam into systems of pipes which branch over them, the heat is chiefly that lately latent in the steam, and which spreads around as soon as the steam, by touching pipes of lower temperature, is condensed to a state of water. The modes of most profitably effecting these purposes have to be considered in a future chapter.

For producing artificial cold, our processes generally involve the circumstance either of a solid changing into a liquid, during which it absorbs, and hides in its new constitution much of the heat previously sensible in it and in the liquid dissolving it, or of a liquid changing into vapour, during which heat equally becomes latent. Thus by dissolving a salt, nitre for instance, in water, we obtain a solution very cold.

In India the common mode of cooling wine for table is to surround the bottles with nitre thus melting ; and the water of the solution being evaporated again before next day, the salt is left ready for use as before. Such is the mutual attraction of water and many salts, that they readily combine with form of liquid, even when the water is used in the solid state of ice ; and as both the water and the salt then make heat latent, the fall of temperature is very great. Thus common salt and snow mixed, dissolve into liquid brine 37° colder than freezing water, or 5° below the zero of Fahrenheit.

The following is a short table of easily procured freezing mixtures :

Frigorific Mixtures.

Substances mixed.		Thermometer sinks.
Common salt..............	1 part	From any temperature to 5° below zero.
Snow, or pounded ice...	2 —	
Common salt..............	5 —	From any temperature to 25° below zero.
Snow or ice	12 —	
Nitrate or ammonia	5 —	
Snow......................	3 —	From 32° above, to 23° below zero.
Diluted sulphuric acid ...	2 —	
Fused potass..............	4 —	From 32° above, to 51° below zero.
Snow....,.................	8 —	
Nitrate of ammonia......	1 —	From 50° to 4° above zero.
Water	1 —	
Sulphate of soda..........	8 —	From 50° to 0° or zero.
Muriatic acid	5 —	

We have already described under other heads the frigorific effect of evaporating in a vacuum or in the air, and of the operation of condensing a gas to squeeze the heat out of it before letting it expand again to a great volume.

For any given substance, the changes of state from solid to liquid, and from liquid to air, happen, under similar circumstances, so precisely at the same temperature, that they mark fixed points in a general scale of temperature, and enable us to regulate and compare our various thermometers. (See Analysis, p. 6.)

As we can neither weigh heat, nor measure its bulk, nor see it, and as, even if our sense of touch were a correct judge in the matter, which it is not, we dare not touch things that are very hot or cold, some other means was wanted for estimating the presence in bodies of this very subtile principle:—and a means has been found in the measurement of its most obvious and constant

effect, namely, that dilatation or expansion of
bodies which again ceases when the heat is withdrawn. Any substance so circumstanced as to
allow this expansion to be accurately measured,
becomes to us a *thermometer* or *measure of heat.*

In *solid* substances, the direct expansion by
heat is so small as to be seen or measured with
difficulty. In *airs,* again, the expansion is very
extensive, but there is the objection that in
any apparatus yet contrived, which will allow their
expansion completely to appear, they cannot be
protected from the varying pressure of the atmosphere—an influence which affects their volume
even more than common changes of temperature.
But *liquids* are free from both disadvantages, and
when placed in a glass bulb, as *a,* having a long
neck or stalk *a b* proceeding from it, into which
the liquid may rise when expanded by heat, to be
measured, they form the most generally con-

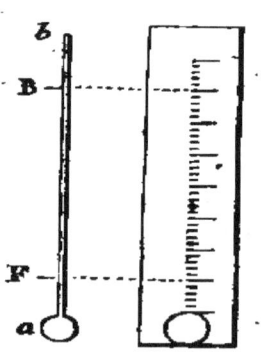

venient of thermometers.
Then, among liquids, mer-
cury stands, on several accounts, singularly pre-emi-
nent : in it the range of temperature between freezing
and boiling reaches a higher
point than in any other liquid, and a lower than in
all others except alcohol ; its
little capacity for heat, and
ready conducting power,
cause it to be very quickly affected by change of
temperature ; its expansion is singularly equable

for equal increase of heat through the important middle part of the scale, which includes the common temperatures on earth, or from the freezing to the boiling heat of water ; and it is easy to proportion the bulb and the stalk to each other, so that a small difference of temperature shall cause the mercurial column in the stalk to rise or fall very conspicuously.

Now when the important fact was ascertained that ice melts in every case at precisely the same temperature, and that pure water in a metallic vessel, and under a given atmospheric pressure, boils always at the same temperature, it followed that by placing such a thermometer as above described, first in melting ice, and then in boiling water, and marking upon the stalk the two points at which the mercury stood, *viz.* F and B, two fixed or invariable points would be obtained, and the interval between them might be divided on the glass, or on a suitable scale to be attached to the glass, into any convenient number of parts to be called degrees : it followed farther, that by continuing the divisions to any extent both above and below the fixed points, a general scale of temperature would be obtained, with respect to which all thermometers made on the same principle would perfectly agree, although the size of the divisions on the stalks would vary according to the comparative capacities of the bulb and stalk in the different instruments. Our Newton had the honour first to propose the regulating points of freezing and boiling, and they are now universally adopted, but the interval between them has been variously subdivided;—that is to say, there has not been agreement among philosophers as to what should be accounted a degree of heat. In the Centigrade thermometer, which is the most simple, the division is into 100 equal parts; in Reaumur's, which is commonly used in France, it is into 80 parts; and in Fahrenheit's, which is used in England, it is into 180°. In Fahrenheit's, moreover, the freezing point, instead of being called *zero*, as in the others, is called 32°, because the maker chose to begin counting from the lowest heat which he met in Iceland, or 32° below freezing of his scale.—To turn the degrees of any one of these thermometers into degrees of any other, we have only to recollect that 9° of Fahrenheit are equal to 5° of the Centigrade, and to 4° of Reaumur. Therefore, multiplying by 9 and dividing by 5 or 4, or the reverse, adding or subtracting the 32° of Fahrenheit, gives as the result the degree desired.

The bulb of a mercurial thermometer is formed by heating in a lamp to fusion the end of a glass tube, which has a very small and equable bore, and then blowing into the tube until the softened end swells like a soap-bubble to the size desired. The mercury is forced into such a bulb through its long stalk by the pressure of the atmosphere at two efforts. First, a portion of the air originally in the bulb being expelled by warming the bulb, the open end of the stalk is immersed in mercury, and when the air still remaining in the bulb cools and contracts, a little mercury enters. Secondly, this admitted mercury having been made to boil, so as to fill with its vapour instead of the air, the whole capacity of the bulb and tube, on the open end being again immersed in mercury, and the mercurial vapour within being condensed, the atmosphere presses in fresh mercury to fill the whole vacuum. To complete the making of the thermometer, the bulb is again heated to expel so much of the mercury as that when cold the tube shall be about one-third full of it, and then before the heated mercury begins to recede, the end or opening is permanently closed by directing upon it the point of a blow-pipe flame.

Although the *direct* expansion of any solid body by a moderate change of temperature is so inconsiderable as to be with difficulty measured, M. Breguet, of Paris, lately with much ingenuity contrived a thermometer in which the index is moved by the curling or bending of a solid when heated, as when a sheet of damp paper curls on being held before the fire. Having soldered side by side two very small flattened wires of silver and platinum, or of any other metals having different expansibility by heat, he found that all changes of temperature made such compound wires bend to a great extent, the metal most shortened or least lengthened acting like a bow-string to pull the other into the arched form : he then, by giving to a compound wire a spiral or cork-screw form, and fixing the upper end of it to a stand, found that an index like the hand of a watch connected with the lower end was turned completely round by a certain change of tempe-

rature, and that when a circle of degrees were
marked on a plate like a watch face placed below
the index, the indications of the instrument perfectly agreed with those of good mercurial thermometers. Other modifications of the same principle have since been successfully tried, so simplified and reduced in bulk as to be introduced
into the structure of a pocket watch.

Air is a substance on several accounts admirably
adapted to the formation of a thermometer ; for
it has great extent of dilatation from small increase of heat ; it quickly receives impression,
and its dilatation is equal for equal increments of
heat at all temperatures :—but, as already stated,
there is the strong objection that the pressure of
the atmosphere cannot be excluded, without at
the same time confining the air, and affecting its
expansion. Mr. Leslie, however, has used for
particular purposes an air-thermometer, which he
calls the differential thermometer. It consists of

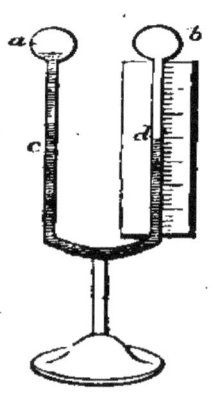

two bulbs a and b, filled with air
and connected by a bent tube $d c$,
containing liquid, the bulbs being
hermetically sealed, so that the
atmosphere cannot affect the air
within. The difference of heat
in one and in the other is marked
by the descending of the liquid in
one of the tubes, d, which has
a scale attached to it. We may
observe that equal divisions or
degrees marked on the scale of
this thermometer cannot mark

equal changes of temperature, as the increasing condensation and resistance of the air in one bulb requires the force overcoming it progressively to increase. If the resistance, on the contrary, were unvarying as in an air-thermometer open to a steady atmosphere, equal extent of motion in the fluid would mark equal increments of heat. An air-thermometer made of a simple bulb and long stalk of semi-transparent porcelain, and containing in its neck melted lead or other fusible metal instead of mercury, and with the mouth downwards, is well adapted for measuring very high temperatures.

Temperatures below that of freezing mercury are usually measured by alcohol, as being a substance which has not yet been frozen: and temperatures higher than of boiling mercury are measured by the expansion of air or of metals, as above described, or by the contraction of pieces of baked clay, which when highly heated lose water and become semivitrified The use of baked clay was proposed by Wedgewood, and the apparatus has been called Wedgewood's *Pyrometer*, or fire-measure. All contrivances for measuring heat may be graduated so as to correspond with the scale adopted for the mercurial thermometer.

It is most interesting, while considering the vast number and importance of the phenomena produced by heat, to observe the degrees in the general scale of temperature at which they take place. In the following table a selection of the facts are classified, the temperatures being all referred to the scale of Fahrenheit's thermometer.

Table of facts connected with the influence of heat corresponding to certain temperatures.

	Fahrenheit.	Wedgewood.
Highest temperature measured	32,277°	240
Chinese porcelain softened	21,357	156
Cast iron thoroughly melted	20,577	150
Greatest heat of a common smith's forge	17,327	125
Flint glass furnace	15,897	114
Stoneware baked in	14,337	102
Welding heat of iron	13,427	90 to 95
Delft ware baked in	6,407	41
Fine gold melts	5,237	32
Settling heat of flint glass	4,847	29
Fine silver melts	4,717	28
Brass melts	3,807	21
Full red heat (*the beginning of Wedgewood's Pyrometer*)	1,077	0
Heat of a common fire	790	
Iron red in the dark	750	
Quicksilver boils	660	
Linseed oil boils	600	

	Fahrenheit.
Lead melts	594°
Sulphur melts	226
Water boils	212
A compound of three parts of tin, five of lead, and eight of bismuth, melts	210
Alcohol boils	174
Bees'-wax melts	142
Ether boils	98
The present medium temperature of the globe	50

	Fahrenheit.
Ice melts	32°
Milk freezes	30
Vinegar freezes................................	28
Strong wine freezes	20
Weak brine freezes ....zero	0
Quicksilver freezes ....below zero	40
Natural temperature observed at Hudson's Bay	50
Greatest artificial cold yet measured	91

There is reason for thinking that the higher temperatures noted in this table appear considerably too high, owing to the insufficiency of the thermometer or pyrometer (Wedgewood's) by which they were estimated.

It is a curious inquiry, suggested by contemplating the preceding table, how much heat may yet remain in bodies at the lowest temperature which we know? No conjecture was hazarded on the subject until Dr. Irvine thought it might be elucidated by comparing the quantity of heat which becomes latent in a body on changing form, with the capacity of the body before and after the change. For instance, with respect to water, he said : as it requires one-tenth more heat to make a certain change in the temperature of water than in that of an equal quantity of ice, it is probable that ice-cold water just contains altogether one-tenth more heat than an equal quantity of ice at the melting point : then as we know the water to contain exactly 140° more heat than the ice, viz. its latent heat, the whole or absolute quantity of heat in it must be ten times 140°, or 1,400°. By applying this reasoning however to other sub-

I

stances than water, it is proved evidently to be
fallacious; and the conclusion follows that we
have as yet no means of solving the question;—
the thermometer no more telling us the absolute
quantity of heat in any body than the rising and
falling of the water-surface in a well tells the total
depth of the well.

From what is said in the last and in preceding
paragraphs, it is evident that the thermometer
gives very limited information with respect to
heat: it merely indicates, in fact, what may be
called the tension of heat in bodies, or the strength
of its tendency to spread from them. Thus it
does not discover that a pound of water takes
thirty times as much heat to raise its temperature
one degree, as a pound of mercury; nor does it
discover the caloric of fluidity absorbed when
bodies change their form, and which indeed is
called latent heat only because hidden from the
thermometer; nor does it tell that there is more
heat in a gallon of water than in a pint; and if
an observer did not make allowance for the increasing rate of expansion in the substance used
as a thermometer, as the temperature increases, he
would believe the increase of heat to be greater
than it is; and, lastly, when a fluid is used as a
thermometer, the expansion observed is only the
excess of the expansion in the fluid over that in the
containing solid, and subject to all the irregularities
of expansion in both substances:—all proving that
the indications of the thermometer, unless inter-
preted by our knowledge of the general laws of
heat, no more disclose the true relations of heat to

bodies, than the money accidentally in a man's pocket tells his rank and riches.

" *Heat by its different relation to different substances has a powerful influence on their chemical combinations.*" (See the Analysis, page 6.)

By observations made and recorded through bygone ages, man has now come to know that the substances constituting the world around him, although appearing to differ in their nature almost to infinity, are yet all made up of a few simple elements variously combined ; and he has discovered that the peculiar relations of these elements to heat,—as their being unequally expanded by it, and their undergoing fusion and vaporization at different temperatures, furnish him with ready means of separating, combining, and new-modifying them to serve to him most useful purposes. Where the primitive savage, looking around on rocks and soils, saw in their diversified aspect almost as little meaning as did the inferior animals which participated with him the shelter of the wood or cave, his son, with penetration sharpened by science, descries at once the treasures of the mine, and aided by heat, whose wonderful energies he has learned to control, pursues through all the Protean disguises of ores and salts and solutions, each of the wished-for substances, until he secures it apart. For instance, in what to his forefathers for thousands of years appeared but a red dross, he knows that there lies concealed the precious iron—king of metals! and soon forcing this in his ardent furnace to assume its metallic form, with implements made of it he afterwards moulds all other bodies to his will : the trees from the forest and the rocks from the quarry, in obedience to these, come to be fashioned by him as if they were of soft clay, and at his command rise into the magnificent structures of palaces and ships, with which the earth is now beautified, and the ocean so thickly covered.—The minute detail of the relations to heat of particular substances forms a great part of the department of science called *chemistry* (a name taken from an Arabic word signifying *fire*) ; but a general review of the subject belongs to this work.

The most common ores of metals are combinations of them with oxygen, carbonic acid, or sulphur, substances all of which are volatilized at much lower temperatures than the metals. Now simple roasting, as it is called, or strongly heating the ores, suffices often to drive away great part of these adjuncts ; and where additional assistance is required, it is obtained by mixing with the ore something which when heated attracts the substance to be expelled more strongly than the metal does. Charcoal, for instance, heated with an oxid-ore, takes the oxygen, and flying off with it as carbonic acid, leaves at the bottom of the furnace or crucible the vivified metal.

Mercury mixed with the dross of a mine, dissolves any particles of gold or of silver existing in it, and the ingredients of the solution may afterwards be obtained separate by mere heating—the mercury passing away as vapour to where it is cooled and again condensed for repeated use, and the more fixed gold or silver remaining pure in its place,—just as in all other distillations, as that of spirit from wine, or of essential oils from water, &c., there is the separation by heat of a more volatile from a less volatile substance. The only difference between what is called drying by heat and distilling is, that in the one case the substance vaporized, being of no use, is allowed to escape or be dissipated in the atmosphere ; while in the other, being the precious part, it is caught and condensed into the liquid form.

A piece of cold charcoal lies in the air for any length of time without change : but if heated to a certain degree, the mutual cohesion of its particles is so weakened, that is, the particles are so repelled and separated from each other, that their attraction for the oxygen in the air around is allowed to operate, and they combine with that oxygen, so as to produce the phenomenon of combustion. The same is true, under similar circumstances, of almost any dry vegetable or animal substance, and of several of the metals.

Nitre, sulphur, and charcoal, while cold, may be mixed together most intimately without any change taking place ; but if the mixture, or any part of it, be heated to a certain degree, the whole explodes with extreme violence, for it is gunpowder. By the change of temperature, and the consequently altered relative attractions of the different substances, a new chemical arrangement of them then takes place with the intense combustion and expansion, which constitute the explosion.

Sea sand and soda may be mixed, and even

ground together, as completely as possible; but
if they remain cold, they remain also merely an
opaque and useless powder : on heating the mixture, however, to diminish the cohesion of the
particles of each substance to those of its own
kind, so that the mutual attractions of the two
substances may come into play, they melt altogether, and unite chemically into the beautiful
compound called glass; a product, than which
art has formed none more admirable—which in
domestic use, for instance, is fashioned into the
brilliant chandelier and lustre, into the sparkling
furniture of the side-board, into the magnificent
mirror-plate, and which, extended across our window openings, admits the light while it repels the
storm.

Perhaps the influence of temperature on chemical union is nowhere more remarkably exhibited,
than in retarding or hastening the decompositions
of dead vegetable and animal substances.   The
functions of life bring into combination, to form
the various textures of organic or living bodies,
chiefly four substances, *viz. carbon* or coal; the
ingredients of water, or *oxygen* and *hydrogen*;
and lastly, *nitrogen*—which substances, when in
the proportions found in such bodies, have
but slight attraction for each other, and all of
which, except the *carbon*, usually exist as airs.
Their connexion, therefore, is easily subverted;
and particularly by a slight change of temperature,
which either so weakens their mutual hold as to
allow new arrangements to be formed, or altogether disengages the more volatile of them.—At a

certain temperature, a solution of sugar (which
consists of the three substances first mentioned,
carbon, oxygen, and hydrogen), undergoes a
change into a spirituous wash, from which spirit
or alcohol may then be obtained by distillation :
but if the heat be continued under certain circumstances, the liquid undergoes a second change, or
new arrangement of constituent particles, and becomes vinegar : under still other circumstances it
undergoes a third change, which is a destructive
decomposition, or rotting, as we call it, and the
oxygen and hydrogen ascend away as airs.   But
sugar, and many similar vegetable compounds, preserved at a low temperature, remain unchanged
for ages.

Again, as regards dead animal substances, we
find that although at a certain, not very elevated,
temperature, they undergo that change in the relations of their elements which we call putrefac-
tion, when nearly their whole substance rises again
to form part of the atmosphere, still at or below
the temperature of freezing water, they remain
unaltered for any length of time.   In the middle
of summer, recently caught salmon, or other fish,
packed in boxes with ice, is conveyed fresh from
the most remote parts of Britain to the capital.
In our warmest weather, any meat or game may
be long preserved in an ice-house.   In Russia,
Canada, and other northern countries, on the setting in of the hard frosts, when the inferior animals
have difficulty in finding food, the inhabitants kill
their winter supply, and store their provender of
frozen flesh or fowl, as in other countries men

store that which is salted or pickled. But the most striking instance of this kind we can adduce is the fact, that on the shore of Siberia, in 1801, in a vast block or island of ice, of which the surface was then more melted than in preceding summers, the carcase of an antediluvian elephant was found, perfectly preserved—an elephant differing materially from those now existing on earth, but its skeleton exactly corresponding with the specimens found deep buried in various countries. The creature was soon discovered by the hungry bears of the district, which were seen tearing off its hairy hide, and feeding on its flesh, as fresh almost as if it had lived yesterday, although it must have been of an era long anterior to that of any existing monument on earth, of human art, or even of human being. Long after it fell from the ice to the sandy beach, and when its tusks had been carried away for sale by a Tungusian fisherman, and its flesh had been nearly devoured, a naturalist who visited it found an ear still perfect, and its long mane, and part of its upper lip, and an eye with the pupil yet distinguishable, which had opened on the glories of a former or younger world! About 30 lbs. weight of its hair, which had been trodden into the sand by the bears while eating the carcase, was collected, and is now preserved in different museums of natural curiosities—some, for instance, in the museum of the London College of Surgeons.

" *Heat has powerful influence also on animated nature, both vegetable and animal.*" (Read the Analysis, page 6.)

As the detail of the relations of heat to particular inanimate substances belongs to the province of chemistry so does the detail of its relations to particular living vegetables and animals belong to the department of Physiology ; but a general review of the subject is required in a treatise on Natural Philosophy.

The influence which heat exerts on inanimate nature, is more immediately and completely perceived by the common mind, than its influence on beings which have life. Thus to all it is obvious, that the contrast between a winter and summer landscape, is owing chiefly to the effect of heat on the water of the landscape ;—that during its absence in winter, there is the dry barren deformity of accumulated ice and snow, covering every thing, the roads impassable, the rivers bound up, perhaps hidden, the air deprived of moisture, and loaded often with powdery drift ;—and that when warmth comes, the living streams again appear, gliding their way, the cascades pour, the rills murmur, the canals once more offer their bosoms to the boats of commerce, the lake and pool again show their level face, reflecting the glories of the heavens, and the genial shower falls upon the bosom of the softened earth, become ready to receive the spade or the ploughshare. Now this change is not at all greater than what happens to a winter tree acted upon by the warmth of spring.—To take another instance from inanimate nature, it may be said with truth, that heat applied to the cold boiler of a steam-engine, is the cause of all its succeeding motions ; of the heaving of its beam and pumps, the opening and shutting of its valves, the turning of its wheels, and its ultimate performance of work, as of spinning, or weaving, or grinding, or propelling vehicles by land and water : but as truly may it be said, that heat coming to a seed which has lain cold for ages, is the cause of its immediate germination and growth ; or coming to a lately frozen tree is the cause of the rising of its sap, the new budding and unfolding of its leaves and blossoms, the ripening of its fruit. And what is true of one seed or tree, is true of the whole of the vegetable creation. When the warm gales of spring have once breathed on the earth, it soon becomes covered, in field and in forest, with its thick garb of green, and soon opening flowers or blossoms every where breathe back again a fragrance to heaven. Among these the heliotrope is seen always turning its beautiful disc to the sun, and many delicate flowers only open their leaves to catch the direct solar ray, but close them often even when a cloud intervenes, and certainly when the chills of night approach. On the sunny side of a hill, or in the sheltered crevice of a rock, or on a garden wall with warm exposure, there may be produced grapes, peaches, and other delicious fruits, which will not grow in situations of an opposite character—all acknowledging heat as the immediate cause, or indispensable condition, of vegetable life.

But among animals, too, the effects of heat are equally remarkable. The dread silence of winter, for instance, is succeeded in spring by one general cry of joy. Aloft in the air the lark is every where carolling ; and in the woods and shrubberies, a thousand little throats are similarly pouring forth their songs of gladness—during the day, •the thrush and blackbird near our dwellings, are heard above the rest, and with the evening comes the sweet nightingale;—for all of which it is the season of love and of exquisite enjoyment. And it is equally so for animal nature generally : in favoured England, for instance, in April and May the whole face of the country resounds with lowings and bleatings and barkings of joy. And even man, the master of the whole, and whose mind embraces all times and places, is far from being insensible to this change of season.   His far-seeing reason of course draws delight from the anticipation of autumn, with its fruits ; and his benevolence rejoices in the happiness observed among all inferior creatures ; but independently of these considerations, on his own frame the returning warmth exerts a direct influence.   In early life, when the natural sensibilities are yet fresh and unaltered by the habits of artificial society, spring to man is always a season of delight.   The eyes brighten, the whole countenance is animated, and the heart feels as if new life were come, and has longings for fresh objects of endearment.   Of those who have passed their early years in the country, or, among the charms of nature, as contrasted with the arts of cities, there are few who, in their morning walks in spring, have not experienced without very definite cause, a kind of tumultuous joy, of which the natural expression would have been, how good the God of nature is to us ! Spring is a time when sleeping sensibility is roused to feel that there lies in nature more than the grosser sense perceives. The heart is then thrilled with sudden extacy, and wakes to aspirations of sweet acknowledgment.

Besides these effects of heat, which are comparavely transient as being connected with the seasons, there are other effects on animated nature of a more permanent character. Certain species of vegetables and animals, by their relation to heat, are confined to certain latitudes or climates; as the orange tree and bird of paradise, to warm regions; the fir tree, and arctic bear, to those that are colder;—and when individuals of either class can support diversity of climate, they acquire a certain character according to the climate,—as seen in the sheep and dogs of the various regions of the earth. In this latter respect there is no instance more interesting than that furnished by the varieties of the human race. Assuming that the whole sprung from one stock, what a contrast is there between the native of Central Africa, of temperate Europe, and of the Polar Zone: between the Negro, the Greek, and the Esquimaux: or again, between the dark slender children of Hindostan, the strongly-knit active Roman or Spaniard, and the taller, ruddy powerful Briton. And in the female sex of the last-named countries, we may remark the gentleness and singular devotedness of the Indian woman, the more commanding dark eye and gesture of the graceful nymph of Italy or Spain, and the happily attempered mixture of these qualities in the fair and much-favoured daughters of Britain.

The very important influence of heat upon the temporary bodily state of animals, becomes an object of much study to the physician : it explains, among many other facts, the connexion of temperature with the rise of fevers and other pestilences, the powerful remedial efficacy of hot and cold bathing, of changes of climate, of regulating the temperature of air breathed by invalids, the protection from clothes, houses, &c.

" *The great natural source of heat is the sun.*"
(See the Analysis, page 6.)

To be assured of this, it is only necessary to think of the comparative temperatures of night and day, of climates and of seasons, and to reflect that the sun is the sole cause of the differences. We need not wonder, then, that, to many savage nations, seeking the source of their life and happiness, the sun has been the object, not only of admiration, but of worship.

The heat comes from the sun with his light. If a sun-beam enter by a small opening an apartment otherwise closed and dark, it illuminates intensely the spot or object on which it first falls, and its light being then scattered around, all the objects in the room become feebly visible. Again, a cold thermometer, held to receive the direct ray, rises much ; while in any other situation it is less affected : proving the heat to be like the light, widely diffused, and so to lose proportionately of its intensity. Light passes from the sun to the earth in about eight minutes of time, as will be fully explained in a future chapter ; and there is every reason to conclude that heat travels at the same rate.

Human art can gather the sun-beams together, and by the intense heat produced in the focus of their meeting, furnishes another proof that the sun is the great source of heat. A pane of glass in a window, or a small mirror, will reflect the sun's ray so as to offend an eye receiving it at a distance of miles—as may be observed soon after the rising, or before the setting of the sun, when his ray is nearly horizontal,—and the heat accompanies the ray, for by many such mirrors directed towards one point, a combustible object placed there would be inflamed. Archimedes set fire to the Roman ships by sun-beams, returned from many points to one, his god-like genius thus rivalling by natural means, the supposed feats of fabled Jupiter with his thunderbolts. Again, when the light of a broad sun-beam is made by a convex glass or lens to converge to one point or focus, the concentrated heat is also there—for a piece of metal held in the focus drops like melting wax : and if the glass be purposely moved, its focus will pierce through the most obdurate substances, as red hot wire pierces through paper or wood. A hunter on his hill, and travelling hordes on the plains, often conveniently light their fires at the sun himself, by directing his energies through a burning glass.

The direct ray of the sun, simply received into a box which is covered with glass to exclude the cold air, and is lined with charcoal or burned cork to absorb heat, and to prevent the escape of heat once received, will raise a thermometer in the box to the temperature of 230° degress of Fahrenheit, a temperature considerably above that of boiling water. And the experiment succeeds in any part of the earth where there is a clear atmosphere, and where the sun attains considerable apparent altitude. We see therefore that a solar oven might in some cases be used. In operating with the apparatus suggested by the author, and described at page 95, for distilling water by the heat of the sun, the vessel intended to absorb the heat, and to act as the still, should be enclosed in a case lined and covered as above described.

Reflecting on such facts as now recorded, and on the globular form and the motions of our earth, we have a measure of the differences of climate and of season that should be found upon the earth. It is evident that the part of the globe turned directly to the sun, receives his rays as abundantly as if it were a perfect plane, similarly facing him, while on parts, which, as viewed from the sun, would be called the sides of the globe, with the increasing obliquity of aspect, an equal breadth or quantity of rays is spread over a larger and a larger surface ; and at the very edge the light passes level with the surface, and altogether without touching. The sunny side of many a steep hill in England, receives the sun's rays in summer as perpendicularly as the plains about the equator ; and such hillside is not heated like these plains, only because the air over it is colder—just as mountain tops, even at the equator, owing to the rarified and therefore cold air around them, remain for ever hooded in snow. In England, at the time of the equinoxes, a level plain receives only about half as much of the sun's light and heat as an equal extent of level surface near the equator ; and in the short days of winter it receives considerably less than a third of its summer allowance.

There are few contrasts in nature more striking than some of the consequences of different intensity of the sun's influence :—that, for instance, of the inhabitants of India, at mid-day, in the hot season, with the thermometer at 120°, running to the shade of their bungalows, darkening their windows, hanging wetted mats upon the walls and roofs, and sprinkling the floors, fanning themselves with ever-moving punkas, and feeling the slightest covering or exertion too much—while, on the other hand, the dwellers in Greenland, with the thermometer below zero, are loaded with furs, and are seeking the direct sunshine or heat from a fire, as their life and comfort. Again, there is the contrast observed on passing, as the author once did, in ten days, from such a paradise as Rio de Janeiro, with all its vegetable riches, to Tristan da Cunha, and the Isle of Desolation in the Southern Ocean, which exhibit only cold and naked rocks ; but yet where the scene was swarming with its appropriate inhabitants—the sea with seals, and the air with clouds of sea fowl, playing over the never-resting waves like flakes of eddying snow. Were a person for a moment to doubt whether the sun be the real cause of such differences, and of certain creatures being found only in certain zones of the earth, let him reflect on the extraordinary migration of animals, which have their home not in any fixed region, but wherever the sun has for a time a particular degree of influence, and which accordingly follow the sun in the changes of season. We have the swallow in such numbers, coming to visit the British isles in the spring, to play over our woods and waters, in pursuit of the insects which the heat then breeds in the air,— welcome harbingers of the coming summer and its riches; and in autumn the same creatures are seen congregating on our shores, to wing their flight back in united multitudes to more southern countries, where, in turn, there is a temperate influence of the sun. The same season brings to England the nightingale, and makes our woodlands resound with the note of the cuckoo. In the waters of our bays and coasts, again, there appear with the seasons the vast shoals of fish, as the herring and mackarel, which prove such abundant food for millions of human beings; and the salmon, at stated times, penetrates from the ocean far up the mountain streams, to deposit its spawn for future supply;—all by their movements, contributing to the harmonious and beneficent system of the universe.

With respect to the sun as a source of heat, there have been two opinions among philosophers; one class believing that the sun is an intensely heated mass, which radiates its heat and light around, like a mass of intensely heated iron: and another class holding that heat is merely an affection or state of an ethereal fluid, which occupies all space, as sound is an affection or motion of air, and that the sun may produce the pheno-

K

mena of light and heat without waste of its temperature or substance, as a bell may without waste continue to produce sound : holding farther, that the sun below its luminous atmosphere may be habitable even by such animals as live on this earth. Those who take the first view, are awakened to the dread contemplation of a universe carrying in itself, if its laws remain constant, the seeds of its certain decay, or at least of great periodical revolutions : the others may view the universe as destined to last nearly unchanged, until a new act of the will of its Creator shall again alter or destroy it.

Of one fact there can be no doubt, *viz.* that the present temperature of the earth is much lower than the temperature in remote past time. The rocks called primitive, as granite and gneiss, constituting the interiors of our great mountain masses, and the substrata of our plains, bear evident marks of having been at one period in a molten state, from which they have been solidified by a very gradual cooling ; and even the whole mass of the earth at some time must have been so fluid or soft, as, in obedience to gravity, to have assumed its rounded form, and in obedience to the centrifugal force of its whirling, to have bulged out, at its great circumference or equator, the seventeen - miles which its equatorial diameter exceeds the polar ; the same, by the bye, in degrees corresponding to the various speed of rotation, being true of all the other planets belonging to the solar system. Again, while in excavating below the surface of the globe, or in examining its structure as exposed to view by volcanic or other convulsions, men encounter in very many situations a thickness of more than a mile, of the wreck and remains of former states of the world—as on digging eighty feet under vineyards near Mount Vesuvius, they encounter the buried cities of Herculaneum and Pompeii—they further discover that the animal and vegetable remains buried, without number, in the present cold climates of the earth, and evidently resting near where the creatures lived, are all of kinds now inhabiting only the warmer or tropical regions. Lastly, in the operations of mining, the deeper men go, the higher they find the temperature to be, at the rate of a degree for about 200 feet of descent; which fact, as heat tends to equable diffusion, proves both that the central heat of our earth must have had another source than a radiation from the sun of the present intensity; and that the surface of the earth is now radiating away more heat than it receives from the sun. The conclusion then follows, that the temperature of the world is still falling, although perhaps so slowly that a change may not be detected even within centuries. Possibly in very remote antiquity that may have been true which the early Greeks erroneously thought true in their day, *viz.* that the equator of the earth, by reason of its great heat, was a barrier impassable by man between the northern and southern hemispheres.

" *Electricity a source of heat.*" (See the Analysis.)

This subject can only be satisfactorily entered upon in the chapter devoted exclusively to elec-

tricity, and is therefore deferred.   Suffice it here
to say, that while an electrical discharge or current
passes from one situation to another, the substance
serving as a conductor is often heated, melted, or
dissipated, in such a manner as to make it doubtful
whether we possess any more powerful means of
producing these effects.   We may remark, too,
that in certain cases of the electrical current, the
heat is accompanied by as intense a light as art
can exhibit.

" *Combustion and other chemical actions as sources
    of heat.*"   (See Analysis, page 6.)

Of the phenomena of nature there is perhaps
none which to the uninstructed appears so inex-
plicable and so wonderful as that of *fire* or *combustion*—whether contemplated in its beauty or in
its terrors.   Fire is seen in its beauty when used
by man for his domestic purposes, as when it
blazes cheerfully over his parlour hearth, or beams
around its steady light from his lamps and chandeliers.   It is seen again in its terrors, when
spreading by accident from some focus, it enve-
lopes in sudden flame the draperies and other
furniture of an apartment ; or when breaking from
a first apartment it rages through a whole habitation, consuming as its food and carrying in its
long flames to the sky, almost every thing save
the stone walls, left as a blackened skeleton ; or
again, when still wider spread, it is at the same
dread moment, devouring with deafening uproar a
whole town or a forest :—nay, it is terrible fire
labouring within the bowels of the earth, which

first prepares and then urges up to heaven the volcanic eruption of flame and red-hot rocks, during which the region around often quakes and is uptorn, with demolition of its cities into sudden tombs of the inhabitants, with change in the course of its rivers, with conversion of its plains into lakes, or of its lake-beds into dry land. Fire appears terrible also in the meteors of night ; and worse than terrible when, intentionally lighted by human hands, it bursts from the cannon's womb to produce the carnage of the battle. Fire among many nations of antiquity was regarded with awe and holy reverence, the sun himself being honoured chiefly as its concentration or supposed abode. Then there were sacred fires in many of the temples, and fire was used to complete the splendour of the most august ceremonies. But, more remarkable still, Moses, a worshipper of the one true God, has recorded of the *Burning Bush*, and of burnt offerings made to that God : and at the present day, in many Christian churches, there are ever-burning lamps and frequent magnificent illuminations. Now this wondrous principle of *fire*, which when the savage man first saw it spreading perhaps after the thunderclap or the rubbing of forest branches in a storm, so as to threaten universal destruction, he so naturally accounted the demon, if not the God of nature,—this principle man's art has now tamed to be a most obedient, and by far the most useful of all his servants. *Fire,* being in truth, but a concentration of the element of *heat,* which in its tranquil and invisible diffusion we have already contemplated as the beneficent life or soul of the universe—the cause of seasons and climates, and of all the changes or activity which distinguish a living world from a dead and frozen mass; man, by acquiring command over it, can command heat when and where he wills, and thus truly becomes in a second degree the ruler of nature. Fire in man's service may be figured as a legion of spirits to whom no labour is difficult, and who in any particular case have power or magnitude exactly proportioned to the quantity of food or fuel afforded; of whom, moreover, man can at any moment conjure up one or many by the magic stroke of his flint and steel. In every private dwelling he has of these fiery spirits as domestic servants—in the kitchen and in the parlour. In his manufactories they are melting glass for him, and reducing ores, and boiling and evaporating for a hundred purposes. But it is chiefly while chained to the steam-engine, that they shew their miraculous powers :—as when, putting forth a giant's strength, they heave a river from the bottom of a mine, or urge a vast ship through the winter storm ; or when in nice dexterity equalling, if not surpassing, what human hands can effect, they twist the silken or cotton threads, and weave them into most delicate fabrics. Men now grown familiar with such prodigies, have almost ceased to be moved by them ; but few persons can resist a feeling of wonder and admiration when chemistry, in its progress of discovery, every now and then calls forth the hidden spirit of combustion in some new or less familiar guise :—for instance, when a piece of iron wire lighted as a taper in oxigen gas, burns with such resplendent brilliancy ;—or when phosphorus similarly placed, throws around its overpowering flood of flame ;—or when small portions of the metal called potassium, being cast upon the surface of water, become as beads of most intense light running about there, and crossing as in a merry dance ;—or, lastly, when flames produced from particular substances are seen rising deeptinged with most vivid and beautiful colours.

Singularly interesting then, to philosophers, as in such particulars the phenomenon of combustion must always have appeared, one may wonder that its true nature could remain to them so long a mystery ; but until the admirable researches of Davy, made only a few years ago, their conjectures had scarcely approached the truth. An opinion long prevailed, that in every combustible substance there was present a certain quantity of a something denominated *phlogiston,* which on being disengaged or separated, became obvious to human sense as light and heat. The white oxid of zinc, for instance, named the flowers of zinc, and into which the metal is changed by burning, was supposed to be the metal deprived of its phlogiston ; and when the metal again appeared, on this oxid being heated with charcoal, it was supposed simply to have recovered phlogiston from the charcoal. The illustrious Lavoisier had the merit of most clearly disproving this hypothesis, by shewing, for instance, that the flowers of zinc were heavier than the piece of metal from

which they were produced, by the exact weight
of the oxygen gas, which disappeared in the combustion, &c. ; and he shewed further, that in this
and many other cases, combustion was merely the
act of two substances combining chemically ; but
he fell into an error almost as great as that which
he overthrew, by supposing that oxygen had always
to be one of the combining substances, and that
the heat and light given out in every case had been
previously latent in that oxygen.

When Sir Humphrey Davy began his labours
on the subject, than which labours there is not
perhaps on record a more perfect specimen of
truly scientific research, it was already known that
bodies when compressed or by any means reduced
in bulk, generally gave out a part of their heat, as
—when air condensed in the match-syringe lights
tinder,—or when water and sulphuric acid uniting
into a compound of smaller volume than the separate ingredients become very hot,—or when
water poured upon quick-lime to slake it, and
becoming solid with it, produces heat sufficient
to inflame wood, as has been fatally proved by
the burning of many lime-loaded ships ;—it
being evident, moreover, that the heat produced during chemical unions depended more
upon the energy of the action which united the
substances than upon the change of volume produced.

Farther, it was known that any substance having
its temperature raised, by whatever means, to 800°
or more of Fahrenheit's thermometer, became incandescent or luminous,—as when iron, or stone,

or any substance not dissipated by heat is placed in a common fire;—in the first degree the substance being said to be red-hot, and at higher temperatures to be white-hot.

Now, out of these two truths Davy constructed his explanation. He asserted that in any case, combustion is merely the appearance produced when substances, which have perhaps still stronger attraction for each other than quick-lime and water, are combining chemically, so as to become heated at least to the degree of incandescence. During the phenomenon there is not, as was formerly supposed, something altogether consumed or destroyed, or something called *phlogiston* escaping : the substances concerned are but assuming a new form or arrangement. Thus if a piece of charcoal be enclosed in a glass vessel filled with air, and of which the mouth dips into a liquid to confine the air, and if the charcoal be then heated to a certain degree, by means of a burning-glass or otherwise, the cohesion of its particles gives way to their attraction for the oxygen of the air around them, and they immediately begin to combine with the air so energetically as to produce a heat still much greater, accompanied by the light or incandescence of combustion. The charcoal, under these circumstances, soon entirely disappears, or is dissolved in the air, as sugar may be dissolved in water ; but if the air be afterwards weighed, it is found to have gained in its weight the exact weight of the charcoal which has disappeared ; and a chemist can again separate the charcoal from the air, and use either for any purpose as before.　In like manner, if a piece of iron wire be heated at one end, which is then plunged into a jar of oxygen gas, it will burn as a most brilliant taper, and will gradually fall in the form of oxidized drops, or scales of iron, to the bottom of the vessel.　Now during this process the quantity of oxygen will be diminished, but if the scales mentioned be collected, they will be found to weigh just as much more than the original wire expended, as there is of oxygen lost or combined with them.　A chemist can separate this iron and oxygen, and exhibit them apart as before, without change.　Again, if iron and sulphur in certain proportions be heated together, they unite with vivid combustion, but the product weighs exactly as much as the original ingredients.

While every instance of combustion is thus only a case of chemical union, going on with such intensity of action as to produce incandescence, still, according to the nature of the substances combining, the appearance produced varies much. It may be, for instance, with flame or without flame.　The great combining substance in nature, that is to say, the most universally distributed, is oxygen, of which the name is now become familiar even to the ears of the unlearned. It forms four-fifths of the substance of water and one-fifth of our atmosphere, being on the latter account present every where, and ready to unite itself with any matter exposed to it at the necessary temperature. Now of substances burning in air, those which are originally aeriform, as coal gas, or which on being heated are vaporized or rendered aeriform

before the union takes place, as oil or wax, assume
the appearance of flame; *viz.* the aeriform particles usually invisible are raised to the incandescent temperature; but when the substance
combining with the oxygen remains solid, while
its particles are gradually lifted away by the
oxygen acting only at the surface of their mass, it
appears during the whole time only as a red-hot
stone. The latter is the case of charcoal, coke,
Welch stone-coal, &c., while in the case of wood,
common coal, &c., a greater or less portion of the
inflammable matter is by the heat of the combustion converted into vapour, and produces the
beautiful appearance of flame.

Of the substances called combustible, and thus
called because they combine with oxygen so energetically as to become incandescent, there are
only a few which will begin to unite or burn
at the common temperature of our globe, the
others requiring to be at some higher and peculiar temperature. Thus phosphorus begins to
burn at 150°, sulphur at 550°, charcoal at 750°,
hydrogen at 800°, &c.; it appearing that up to
these temperatures the attraction of the atoms of
the substances among themselves is sufficient to
resist the other attraction, or that of oxygen. But
when the combustion once begins, the temperature,
from the effect of the combustion itself, rises instantly much beyond the degree necessary for the
commencement of the process. Oxygen and hydrogen, which begin to burn or combine at 800°,
produce a flame of as intense heat as human art
can excite.

On the circumstance that bodies require to have a certain preparatory temperature before beginning thus to combine with oxygen, depend many important facts in nature and art. Hence the safety with which most combustibles may be exposed at ordinary temperatures to the contact of atmospheric air : otherwise coal, wood, &c. in the moment of being exposed to the air would catch fire, as really happens to phosphorated hydrogen gas; or to the metal called potassium, even when thrown into cold water, the metal attracting the oxygen from the water instantly, and with intense combustion. If a fire or flame be so small that it does not produce heat enough to maintain the inflaming temperature of the substance, the combustion will soon be extinguished. Thus a common coal fire, if not watched by gathering together the remnants to reduce the surface of wasteful radiation, will be extinguished long before the fuel is all expended :—but not so with a fire of wood or of paper, which substances burn more readily than coal. The Welch stone-coal can only be made to burn in very large masses, or when mixed with a more inflammable coal or other fuel. A substance placed in pure oxygen gas burns with much greater intensity, and will begin burning at a lower temperature than if placed in atmospheric air, which contains only one-fifth of oxygen and four-fifths of another substance, nitrogen, which does not aid the combustion,—because the nitrogen, by absorbing much of the heat of the combustion, lowers the temperature. Iron wire will burn as a taper in oxygen, but not in common air; and a common taper or flaming piece of wood just extinguished by blowing on it, will immediately be rekindled if placed in oxygen. Again, a lamp with a very small wick, as of one thread, and producing therefore very little heat, will not burn in cold weather, and at any time will be extinguished by a foreign body, brought near it so as to cool it,—a small metallic nob, for instance, presented to it on the end of a wire, or a metallic ring let down over it; but if the ball or ring be hot, the effect will not follow. By more powerful refrigerating processes even a considerable lamp or candle may be put out. These discoveries led Davy to the construction of his miner's safety lamp, which is merely a lamp surrounded by a wire gauze, of which the meashes are of such size that a flame of the gas attempting to pass through is so cooled by the heat-absorbing and heat-conducting power of the metal, as to be extinguished. A wire gauze gradually let down upon any common flame, annihilates the part of it which should appear above the gauze; but the combustible vapour passing invisibly through the gauze may be lighted afresh on its upper side. Oxygen and hydrogen, which are the constituents of water, when uniting, produce such intense heat that the momentary expansion of the newly formed water—then in the state of steam, is such as to constitute a violent explosion : and when meeting jets of the two gases at a certain point allow a continued flame to be formed, the most refractory substances melt in it like wax in a common taper,—yet these gases may be kept mixed toge-

ther in the cold reservoir of a condensed air blow-
pipe without combining, and when they are set
on fire issuing as a jet from a small opening, the
flame does not travel inwards through the opening as might be feared, because it is cooled by the
metal of the orifice.

While solid bodies become very visible or incandescent at about 1,000° of Fahrenheit, airs,
owing to their tenuity of condition, require to be
heated much farther before they take on the vivid
appearance of flame; and airs of light atoms, like
hydrogen, require to be heated still more than
heavier airs. Thus a wire held in the pale blue
flame of pure hydrogen, becomes much more luminous than the flame itself; and the flame of
mixed oxygen and hydrogen escaping from a very
minute orifice in a glass tube, may itself be scarcely
visible, while the extremity of the tube heated by
it becomes like a brilliant star. Hence the light
of many flames may be increased by placing a
wire gauze or other solid body in the flame. Consideration of this subject enables us to explain
why common coal gas, which consists of hydrogen
holding a quantity of carbon in solution, gives
in burning a stronger light than pure hydrogen,
and why oil gas, which contains about twice as
much carbon as the coal gas, gives also about
twice as much light :——for it appears that the atmospheric air, which first mixes with these gases
as they issue to burn, is sufficient to combine with
all their hydrogen (which it most strongly attracts),
but not at the same time with all their carbon ;
the particles of the carbon therefore are sepa-

rated or precipitated in the flame, and become so many solid particles most intensely heated and luminous; and afterwards when they have ascended a little higher, they meet with new, oxygen and burn in their turn, giving a second dose of light. That this decomposition of the gas really occurs is proved by placing a wire gauze in the flame, when we find that if held near the middle of the flame, it is immediately loaded with the particles of charcoal separated there, and cooled by it so as to cohere; while if held at the bottom of the flame where the carbon is not yet separated, it retains none, and if held at the top of the flame, where they are already burned, it similarly retains none. A candle or lamp is said to smoke when the heat produced by it is not sufficient to effect the total combustion of the carbon which rises in its flame.

When oxygen mixed with certain of the inflammable gases or vapours is raised to a temperature even considerably below that of common burning or explosion, a union still takes place, but very slowly, so that the temperature never rises to that necessary to exhibit flame. This phenomenon has been called invisible combustion. It is remarkably exemplified on plunging platinum or gold wire moderately heated into such a mixture : the combination then goes on in the immediate vicinity of the hot wire; and although without flame, still with sufficient disengagement of heat to maintain the wire in an incandescent or luminous state, as long as there are gases left to combine. Thus the vapour always arising at a common temperature from the mouth of a phial of ether (ether com-sists chiefly of hydrogen and carbon), if made to pass through a coil of heated platinum wire, will, while by this slow combustion, combining with the oxygen of the air around it, give out heat enough to keep the wire so luminous as to serve as a little lamp by which to read from the dial-plate of a watch through the night. A beautiful modification of this principle has been adopted in the miners' safety lamp ; and when the air of the mine is too impure to maintain the flame, it still suffices thus to produce a continued light from the incandescent metal.

" Fuel. "

HEAT being, in the sense already explained, the life of the universe, and man having command over nature chiefly by his power of controlling heat, which power again comes to him with the ability to produce combustion, it is of great interest to inquire what substances he can most easily procure as food for combustion, or *fuel*, as it is called, and how these may be most advantageously employed. To speak on this subject at all fully in reference to the various arts of life would be to compose an extensive work, but an interesting sketch may be comprised within narrow limits.

Although there are a great number of substances, which in the act of their chemical union occasion the heat and light which constitute combustion, still by far the greater part of these, in an uncombined state, are so sparingly distributed in nature, and are therefore procurable with such difficulty, that heat obtained by sacrificing them would be much too expensive to be within common means. Providence however has willed that the elementary substance in nature which has the most energetic attraction for almost all other substances, and which therefore produces in uniting with them the most intense heat, is also the most universally distributed of all. This substance is *oxygen.* It forms part of our atmosphere, and therefore penetrates, and is present wherever man can exist or breathe, offering itself at once to his service. Then for the purpose of combining with the oxygen, there are chiefly two other substances also very widely scattered, and therefore easily procurable and cheap. These are carbon and hydrogen, the great materials of all vegetable bodies, and therefore of our forest trees, and of coal beds, which seem to be the remains of antediluvian forests. Carbon is found nearly alone in hard coal, but it is united with a large proportion of hydrogen in caking coal, wood, wax, resins, tallow, and oils. The gases used for illumination are merely hydrogen, holding certain quantities of carbon in solution; and all bodies which burn with flame give out such gases in the act of combustion. In the great mass of the earth as known to man, the stones, earths, and water, forming its surface, are already combinations of oxygen with other substances, and are therefore not in a state to produce fresh combustion; but carbon and hydrogen, by various processes of vegetable and animal life, are in numberless si-

tuations becoming accumulated, so as to be fit for
fuel :—as by other processes the atmosphere is
always preserved with its due proportion of
oxygen.

The name fuel is given only to the substances
which combine with oxygen, and not to the oxygen
itself, probably because the former being solid or
liquid, and therefore more obvious to sense, were
known as producers of combustion long before the
existence of the aeriform ingredient was even suspected.

Oils, fat, wax, &c. being or becoming in their
combustion aeriform, exhibit the appearance of
flame, as already explained, and hence are chiefly
used for the purpose of giving light. Wood,
again, and coal, are more frequently used for mere
heating. But the chemist's lamp for distilling and
evaporating, his common blow-pipe for directing
the point of a flame upon any substance to melt
it, and his condensed-air blow-pipe, whose flame
of oxygen and hydrogen is capable of melting
the most refractory substances, prove that it is
chiefly the expense of the former kinds of fuel
which has nearly limited them to the office of
light-giving. Lately an important application of
oil or fat as heat-giving fuel has been made in a
general cooking apparatus, which promises to effect
a considerable diminution of house-keeping expense.

Wood was the common fuel of the early world
when coal mines were not yet known, and still in
many countries it is so abundant as to be the
cheapest fuel. Charcoal is the name given to

what remains of wood after it has been heated in a close place, during which operation the hydrogen and other minor ingredients are driven away in the form of vapour. Charcoal is nearly pure carbon. Coke, again, is the carbon obtained by a similar preparation of coal. The wood and coal, if similarly heated in the air, would burn or combine with the oxygen of the air; but heated in a vessel or place which excludes air, they merely give out their more volatile parts.

Good coal, where it abounds, is now for ordinary purposes by much the cheapest kind of fuel; and since within a few years men have learned to obtain from it separately, and to use instead of oil and wax, its illuminating gas, *viz.* its hydrogen, holding in solution a little carbon, it has become doubly precious to them. A person reflecting that heat is the magic power which vivifies nature, and that coal is what best gives heat for the endless purposes of human society, cannot without admiration think of the rich stores of coal which exist treasured up in the bowels of the earth for man's use. And Britain, in this respect, is singularly favoured. Her coal mines are in effect mines of labour or power vastly more precious than the gold and silver mines of Peru, for they may be said to produce abundantly every thing which labour and ingenuity can produce, and they have essentially contributed to make her mistress of the industry and commerce of the earth. Britain has become to the civilized world around, nearly what an ordinary town is to the rural district in which it stands, and of this vast and glorious city the mines in question are the coal-cellars, stored at the present rate of consumption for about 1,000 years; a supply which, as coming improvements in the arts of life will naturally bring economy of fuel, or substitution of other means to effect similar purposes,—may be regarded as exhaustless.

Coal, we can scarcely doubt, is the remains of antediluvian forests, swept together during convulsions of nature into deep vallies, and there afterwards compressed and solidified by superincumbent deposits of earthy matters, these deposits being probably aided in their operation by heat. In many coal beds the trees of former times yet retain their form, so that their species can be easily distinguished, and there are buried among them other vegetable and animal remains of contemporaneous inhabitants of the earth. Coal is found of different qualities. In some places it is almost unmixed carbon, and exceedingly solid, as if it had been coked by subterranean heat. Such is the stone-coal of Wales, which in 100 parts contains 97 of pure carbon, and only three of hydrogen and earthy matter. In other places the coal contains hydrogen in nearly as large proportion as wood does, and so combined with part of the carbon as to form the oily or pitchy substances existing in the coal, and which when burning produce flame, and when rising unburned constitute smoke.

The comparative values, as fuel, of different kinds of carbonaceous matter, have been found on experiment to be as in the following tables.

1 lb. of	Melts of ice.
Good coal	90 lb.
Coke	84
Charcoal of wood	95
Wood	32
Peat	19

Lavoisier, in making experiments on combustibles generally, to ascertain the quantities of oxygen expanded, and of heat given out during the combustion of a given quantity of each, obtained the following results:

1 lb. of	Melts of ice.	Takes of oxygen.
Hydrogen gas	370 lbs.	$7\frac{1}{2}$ lbs.
Carburetted hydrogen...	85	4
Olive oil	120	3
Wax......................	110	3
Tallow....................	105	3
Charcoal	95	$2\frac{2}{8}$
Phosphorus	100	$1\frac{1}{8}$
Sulphur	25	1

There are some remarks with respect to the using of common fuel, which seem to demand a place here.

A pound of coke produces nearly as much heat as a pound of coal; but we must remember that a pound of coal gives only three-quarters of a pound of coke, although the latter is more bulky than the former.

It is wasteful to wet fuel, because the moisture in being evaporated carries off with it as latent, and therefore useless heat, a considerable proportion of what the combustion produces. It is a very common prejudice, that the wetting of coal, by making it last longer, is effecting a great saving; but while, in truth, it restrains the combustion, and for a time makes a bad fire, it also wastes the heat.

Coal containing much hydrogen, as all flaming coal does, is used wastefully when any of the hydrogen escapes without burning; for, first, the great heat which the combustion of such hydrogen would produce is not obtained; and, secondly, the hydrogen, while becoming gas, absorbs still more heat into the latent state than an equal weight of water would. Now the smoke of a fire is the hydrogen of the coal rising in combination with a portion of carbon. We see therefore that by destroying or burning smoke, we not only prevent a nuisance, but effect a great saving. The reason that common fires give out so much smoke is, either that the smoke, or what we shall call the vaporized pitch, is not sufficiently heated to burn, or that the air mixed with it as it ascends in the chimney, has already, while passing through the fire, been deprived of its free oxygen. If the pitch be very much heated, its ingredients assume a new arrangement, becoming transparent, and constituting the common coal gas of our lamps; but at lower temperatures, the pitch is seen jetting as a dense smoke, from cracks or openings in the coal—a smoke, however, which immediately becomes a brilliant flame if lighted by a piece of burning paper or the approximation of the combustion. The alternate bursting out and extinction of these burning jets of pitchy vapour, contribute to render a common fire an object so lively, and of such agreeable contemplation in the winter evenings. When coal is first thrown upon a fire, a great quantity of vaporized pitch escapes as a dense cold smoke,—too cold to burn, and for a time the flame is smothered, or there is none; but as the fresh coal is heated its vapour reproduces the flame as before. In close fire-places, those *viz.* of great boilers, as of steam-engines, brewing, and dyeing apparatus, &c., all the air which enters after the furnace-door is shut, must pass through the grate and the burning fuel lying on it, and there its oxygen is consumed by the red-hot coal before it ascends to where the smoke is. The smoke therefore, however hot, passes away unburnt, unless sometimes, as over foundery furnaces, where the heat is very great indeed, and it burns as a flame or great lamp at the chimney-top on reaching the oxygen of the open atmosphere.

There have been many modes proposed of destroying smoke: one has been to admit, by a suitable opening, a certain quantity of fresh air to the space above the fire, the oxygen of which air may inflame the smoke. At a certain point of time after the addition of fresh fuel, this plan succeeds, and for the moment effects a saving of fuel; but the difficulty of admitting just the quantity of air required to suit the varying demand for it, has not been overcome, and hence from there being no saving on the whole, the plan has been abandoned. When just enough air entered, the flame produced gave so intense a heat as in several cases to have burned or destroyed the parts of valuable boilers exposed to it ; and when, on the contrary, too much air entered, it injuriously cooled the boiler.   The contrivance at present most commonly adopted for burning smoke, is that of Mr. Brunton, *viz.* a circular fire-grate, kept turning like a horizontal wheel, and on which coal is by machinery made to fall in a gradual manner, so as to be uniformly spread over it.   The coal falls so gradually, that although there is generally a little smoke from it, there is never much,—the oxygen which finds entrance, through, and around the grate, being always in quantity the same, and nearly sufficient.   A smoke-consuming fire would be constructed on a perfect principle, in which the fuel were made to burn only at the upper surface of its mass, and so that the pitch and gas disengaged from it, as the heat spread downwards, might have to pass through the burning coals where fresh air were mixing with them ; thus the gas and smoke, being the most inflammable parts, would burn first and be all consumed.   This was the principle proposed in a fire-place suggested by the author for the great brewery of Mr. Meux in his neighbourhood, and tried at the time when attempts were extensively made to abate the nuisance of smoke in towns. The experiment proved the theoretical perfection of the method, and that it would produce a saving of 15 or 20 per cent. on the expenditure of coal ; but before a durable grate of the kind was completed the Welch stone-coal was introduced, which has 97 per cent. of pure carbon, and therefore no pitch to evaporate, and no smoke,—and it was at once adopted there and in many other places. Coal in a deep narrow trough, as *a b c d;* if lighted at its surface *a b,* burns with a lofty flame as if it were the wick of a large lamp; for all the

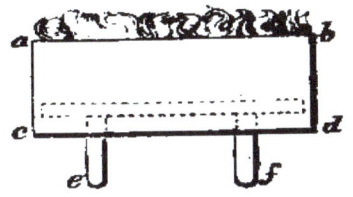

gas given out from the coal below, as that is gradually heated, passes through the burning fuel and becomes a flame. Now, if we suppose many such troughs placed together, with intervals between them, in place of the fire-bars of a common grate or furnace, there would be a perfect no-smoking fire-place. Such was that made on the occasion mentioned; and although flimsy and imperfect, as a mere experimental apparatus, it put beyond a doubt the possibility of accomplishing its object. The reason of the vast saving of fuel by such a grate is, that the smoke, instead of stealing away latent heat—being yet itself the most combustible and precious part of the fuel, gives all its powers and worth to the purpose of the combustion. The coal rested on moveable bottoms in the troughs, and was moved up like the wick of a lamp, by its screw:—the bottoms might be lifted in many ways. The author believes that this construction, simplified as much as possible, will still be adopted for the Newcastle or flaming coal,—the consequences would be so important. The principle has been already extensively introduced for common parlour fires by Mr. Cutler in his stove, which is merely a common grate, having instead of bottom-bars a deep box to hold the coal for a whole day, with a moveable bottom, which lifts the coal up as wanted. From such a fire there is always ascending a long beautiful flame; and much more heat is given out, than from the same quantity of coal burned in the common way : the chimney never requires sweeping, for there is absolutely no smoke, and therefore no soot.

It is evident that if a house or apartment with the air in it, were once warmed to a certain degree, it would for ever retain its temperature, but for the escape of heat through the walls and windows, or with the air from within, whether passing away as necessary ventilation or as waste. A perfect system of heating, therefore, would consist in diminishing as much as possible these causes of loss, with reference both to the expense of the means and the salubrity of the dwelling, and in producing and distributing the heat judiciously. It may be asserted that a fourth part of the fuel generally expended in English houses, if more skilfully used, would better secure comfort and health than all that is now expended. But it does not accord with the character of this general work to enter into minute detail on the subject. Remarks were made upon it in vol. i. in the chapter on " *Pneumatics*," under the head of " warming and ventilating," and more minute information may be obtained from Mr. Tredgold's work, expressly devoted to it.

The consideration of furnaces, blow-pipes, &c. may appear to some so closely connected with our present subject as to demand a place here, but by treating of them we should be encroaching on the province of the chemist, &c. We may state generally, that furnaces are merely arrangements of parts by which coal or other fuel heated to the degree at which it combines rapidly with the oxygen of the atmospheric air, is placed in circumstances favourable to the rapid renewal of the air,—and a common blow-pipe is merely a jet of air thrown from a minute opening into any flame, so as with great precision to direct the point of the flame upon the body to be heated. The sand-bath and water-bath of the chemist are merely means of insuring a more uniform or steady temperature :— a vessel imbedded in sand, so that heat can reach it only through the sand, cannot be very suddenly heated or cooled, because sand is a slow conductor ; and a vessel immersed in boiling water, can never have greater heat than 212°, or the boiling heat of water. For certain purposes, hotter baths, as of high-pressure steam, or of vapour of oil of turpentine, or of boiling whale-oil, have been used. On such subjects, readers may consult works on " chemistry applied to the arts."

" *Condensation and Friction as causes of Heat.*"
(Read the Analysis, page 6.)

A soft iron nail laid upon an anvil, and receiving in rapid succession three or four powerful blows of a hammer, becomes hot enough to light a match, and if longer hammered, will become incandescent or red-hot,—partly from the diminished volume or condensation of the iron, on the principle already explained, and partly from the percussion or friction, in a way not yet well understood, but probably electrical.

In the familiar case of the mutual percussion of flint and steel, small portions of one or both are struck off by the violence of the collision, in a state of white heat, and the particles of the iron burn in passing through the air :—in a vacuum the heated particles are equally produced, but are scarcely visible from this combustion not occuring. In both cases, they suffice to inflame gunpowder, or to light tinder. When the materials are good, the shower of sparks from the sudden blow is most copious and brilliant.

The heat produced by friction alone, without perceivable condensation of the bodies concerned, is exemplified in many facts. Two dry branches kept strongly rubbing against each other by the wind, have sometimes set a wood on fire. Savages light their fires by analogous means. Men warm their cold hands in winter, by rubbing them against each other, or against their coat-sleeves. Again, the axletree of a heavily laden waggon or other carriage, if left without oil, often inflames. The line attached to a whale-harpoon, as it runs over the side of the boat when the huge monster dives after the harpoon has entered his flesh, requires water to be constantly thrown on it to prevent its setting fire to the boat. A cable drawn very rapidly through the hawse-hole by the falling anchor, produces great heat there and smoke. When a magnificent ship is launched from the builder's yard into the deep, and glides along the sloping beams, a dense smoke rises from the points of rubbing contact.

" *The Functions of Animal Life a source of Heat.*" (Read the Analysis, p. 6.)

IT is one of the remarkable facts in nature, that living animal bodies, and to a certain degree living vegetables also, have the property of maintaining in themselves a peculiar temperature, whether surrounded by bodies that are hotter or colder than they. Captain Parry's sailors, during the polar winter, where they were breathing air that could freeze mercury, still had the natural warmth in them of 98° of Fahrenheit; and the inhabitants of India, where the same thermometer stands sometimes at 115° in the shade, have their blood at no higher a temperature.

It was at one time the favourite explanation of this, that animal heat was produced in the lungs, during respiration, from the oxygen then admitted. This oxygen combines with carbon from the blood, and becomes carbonic acid as in combustion, and it was supposed to give out a portion of its latent heat, as in actual combustion; which heat being then spread over the body by the circulating blood, maintained the temperature. We now know that if such a process assist, which it probably does,—for the animal heat has generally a relation to the quantity of oxygen expended in any particular case, and when an animal being already much heated needs no increase, very little oxygen disappears,—still much of the effect is dependent on the influence of the nerves, either

directly, or indirectly through the vital functions
governed by them. Mr. Brodie, in his important experiments upon the subject, found that
although in animals apparently dead from injury
done to the nervous system, he could artificially
continue the action of respiration, with the
usual formation of carbonic acid, still the temperature fell very quickly. The maintenance
of low temperature in an animal immersed in
air hotter than itself, is partly attributable to
the copious perspiration and evaporation which
then take place, and which absorb into the
latent form the excess of heat then existing.
Perspiration, both from the skin and internal
surface of the lungs, occurs generally in proportion to the excess of heat. Dogs and other animals, when much heated, as they cannot throw
off or diminish their natural covering, increase
the evaporating surface by protruding a long humid tongue.

The power in animals of preserving their peculiar temperature has its limits. Intense cold
coming suddenly upon a man who has not sufficient protection, first causes a sensation of pain,
and then brings on an almost irresistible sleepiness,
which if indulged would be fatal. Sir Joseph
Banks having gone on shore one day near the cold
Cape Horn, and being fatigued, was so overcome
by the feeling mentioned, that he intreated his
companions to let him sleep for a little while.
His prayer granted, might have allowed that sleep
to come upon him which ends not—the sleep of
death! as, under similar circumstances, it came

upon so many thousands of the army which Buonaparte led into Russia, and lost there during the disastrous retreat through the snows. Buonaparte's celebrated bulletin allowed that in one night, when the thermometer of Reaumur stood at 19° below zero, 30,000 horses died! Cold in inferior degrees, and longer continued, acting on persons imperfectly protected by clothing, &c. induces a variety of diseases, which destroy more slowly. A great excess of heat, again, may at once excite a fatal apoplexy, and heat in inferior degrees, but long continued, may cause those fevers, &c. which prevail in warm climates, and which are so destructive to strangers in these climates.

Each species of animal has a peculiar temperature natural to it, and in the diversity are found creatures fitted to live in all parts of the earth, what is wanting in internal bodily constitution being found in the admirable protecting covering which nature has provided for them—covering which grows from their bodies, with form of fur or feather, in the exact degree required, and even so as in the same animal to vary with climate and season. Such covering, however, has been denied to man ; but the denial is not one of unkindness : —it is the indication of his superior nature and destinies. Godlike reason was bestowed on man, by which he subjects all nature to his use, and he was left to clothe himself.

The human race is naturally inhabitant of a warm climate, and the paradise described as Adam's first abode, may be said still to exist over vast regions about the equator. There the sun's

influence is strong and uniform, producing a rich
and warm garden, in which human beings, however ignorant of the world which they had come too
inhabit, would have their necessities supplied almost by wishing.   The ripe fruit is there always
hanging from the branches; of clothing there is
required only what moral feelings may dictate, or
what may be supposed to add grace to the form ;
and as shelter from the weather, a few broad leaves
spread on connected reeds, will complete an Indian hut.   The human family, in multiplying and
spreading in all directions from such a centre,
would find, to the east and west, only the lengthened paradise, with slightly varying features of
beauty ; but to the north and south, the changes
of season, which make the bee of high latitudes
lay up its winter store of honey, and send mi-
grating birds from country to country in search of
warmth and food, would also rouse man's energies
to protect himself.   His faculties of foresight and
contrivance would come into play, awakening industry ; and as their fruits, he would soon possess
the knowledge and the arts which secure a happy
existence in all climates, from the equator almost
to the pole.   It is chiefly because man has learned
to produce at will, and to command the wonder-
working principle of heat, that in the rude winter,
which seems the death of nature, he, and other
tropical animals and plants which he protects, do
not in reality perish—even as a canary-bird escaped from its cage, or an infant exposed among
the snow-hills.   By producing heat from his fire,
he obtains a novel and most pleasurable sort of

existence; and in the night, while the dark and freezing winds are howling over his roof, he basks in the presence of his mimic sun, surrounded by his friends and all the delights of society; while in his store-rooms, or in those of merchants at his command, he has the treasured delicacies of every season and clime. He soon becomes aware, too, that the dreary winter, instead of being a curse, is really in many respects a blessing, by arousing from the apathy to which the eternal serenity of a tropical sky so much disposes. In climates where labour and ingenuity must precede enjoyment, every faculty of mind and body is invigorated; and hence the sterner climates form the perfect man. It is in them that the arts and sciences have reached their present advancement, and that the brightest examples have appeared of intellectual and moral excellence.

PART FOURTH.

(Continued.)

SECTION II.—ON LIGHT, or OPTICS.

ANALYSIS OF THE SECTION.

Light is an emanation from the sun and other luminous bodies, becoming less intense as it spreads, and which, by falling on other bodies, and being reflected from them to the eye, renders them visible. It moves with great velocity, and in straight lines where there is no obstacle,—leaving shadows where it cannot fall. It passes readily through some bodies—which are therefore called transparent, but when it enters or leaves their surfaces obliquely, it suffers at them a degree of bending or refraction proportioned to the obliquity. And a beam of white light thus refracted or bent, under certain circumstances, is resolved into beams of all the elementary colours, which however, on being again blended, become the white light as before.

Transparent bodies, as glass, may be made of such form as to cause all the rays which pass through them from any given point to bend and meet again in another point beyond them ;— the body then, because usually in form somewhat resembling a flat bean or lentil, being called a LENS. *And when the light thus proceeding from every point of an object placed before a lens is collected in corresponding points behind it, a perfect image of the object is there produced, to be seen on a white screen placed to receive it, or in the air, by looking towards it in a certain direction. Now the most important optical instruments, and even the living eye, are merely arrangements of parts for producing and viewing such an image under variety of circumstances. When this image is received upon a suitable white surface or screen in a dark room, the arrangement is called, according to minor circumstances, a* CAMERA OBSCURA, *a* MAGIC LANTERN, *or* SOLAR MICROSCOPE. *And the* EYE *itself is, in fact, but a small camera obscura,—of which the pupil is the round opening or window before the lens,—enabling the mind to judge of external objects, by the size, brightness, colour, &c. of the very minute but most perfect images or pictures formed at the back of the eye, on the smooth screen of nerve called the retina. The art of painting aims at producing on a larger scale such a picture, and which when afterwards held before the eye, and reproducing itself in miniature upon the retina, may excite the same impression as the original objects.—When the image beyond a lens, formed as above described, is viewed in the air, by looking at it in a particular direction, then there is exhibited the arrangement of parts constituting the* TELESCOPE, *or* COMMON MICROSCOPE.

Light falling on very smooth or polished surfaces, is reflected so nearly in the order in which it falls, as to appear to the eye as if coming directly from the objects originally emitting it,—and such surfaces are called mirrors. Mirrors may be plane, convex, or concave; and certain forms will produce images by reflexion, just as lenses produce them by refraction; so that there are reflecting telescopes, microscopes, &c., as there are refracting instruments of the same kind. Light, again, falling on bodies of rougher or irregular surface, or which have other peculiarities, is so modified as to produce all those phenomena of colour and varied brightness seen among natural bodies, and giving them their distinctive characters and beauty.

" *Light.*"   (See the Analysis.)

THE phenomena of *light* and *vision* have always been held to constitute a most interesting branch of natural science ; whether in regard to the beauty of light, or its utility. The beauty is seen spread over a varied landscape—among the beds of the flower-gardens, on the spangled meads, in the plumage of birds, in the clouds around the rising and setting sun, in the circles of the rainbow. And the utility may be judged of by the reflexion, that had man been compelled to supply his wants by groping in utter and unchangeable darkness, even if originally created with all the knowledge now existing in the world, he could scarcely have secured his existence for one day. Indeed, the earth without light would have been an unfit abode even for grubs, generated and living always amidst their food. Eternal night would have been universal death. Light, then, while the beauteous garb of nature, clothing the garden and the meadow,—glowing in the ruby— sparkling in the diamond,—is also the absolutely necessary medium of communication between living creatures and the universe around them. The rising sun is what converts the wilderness of darkness which night covered, and which to the young mind, not yet aware of the regularity of nature's changes, is so full of horror, into a visible and lovely paradise.   No wonder, then, if, in early ages of the world, man has often been seen bending the knee before the glorious luminary, and worshipping it as the God of Nature. When a mariner, who has been toiling in midnight gloom and tempest, at last perceives the dawn of day, or even the rising of the moon, the waves seem to him less lofty, the wind is only half- as fierce, sweet hope beams on him with the light of heaven, and brings gladness to his heart.   A man, wherever placed in light, receives by the eye from every object around—from hill and tree, and even a single leaf,—nay, from every point in every object, and at every moment of time, a messenger of light to tell him what is there, and in what condition.   Were he omnipresent, or had he the power of flitting from place to place with the speed of the wind, he could scarcely be more promptly informed. And even in many cases where distance intervenes not, light can impart at once, knowledge which, by any other conceivable means, could come only tediously, or not at all. For example, when the illuminated countenance is revealing the secret workings of the heart, the tongue would in vain try to speak, even in long phrases, what one smile of friendship or affection can in an instant convey; —and had there been no light, man never could have been aware of the miniature worlds of life and activity which, even in a drop of water, the microscope discovers to him; nor could he have formed any idea of the admirable structure belonging to many minute objects. It is light, again, which gives the telegraph, by which men converse from hill to hill, or across an extent of raging sea,—and which pouring upon the eye through the optic tube, brings intelligence of events passing in the remotest regions of space.

" *Emanation from the sun,*" &c. (See the Analysis, page 162.)

The relation of the sun to light is most strikingly marked in the contrast between night and day; as the relation between combustion and light is seen in the brilliancy of an illuminated hall or theatre, as compared with the perfect darkness when the chandeliers are extinguished. In tropical countries, where the sun rises almost perpendicularly, and allows not the long dawn and twilight of temperate latitudes, the change from perfect darkness to the overpowering effulgence of day, is so sudden as to be most impressive. An eye turned to the east has scarcely noted a commencing brightness there, when that brightness has already become a glow; and if clouds be floating near to meet the upward rays, they appear as masses of golden fleece suspended in the sky: a little after the whole atmosphere is bright, and the stream of direct light bending round, makes the lofty mountain-tops shine like burnished pinnacles; then as the stream reaches to still lower and lower levels, the inhabitants of these in succession see the radiant circle first rising above the horizon like a tip of flame, but soon displaying, as in days of Pagan worship, all its breadth and glory,—too bright for the eye to dwell upon. With evening the same appearances recur in a reversed order, ending, as in the morning they began, in complete darkness.

Light emanates also from the stars, but they are so distant as in that respect to be of little importance to this earth. And there are still other transient sources in animal and vegetable nature, and among solar phosphori, but they do not merit particular attention here.

There have been two opinions respecting the nature of light: one, that it consists of extremely minute particles darting all around from the luminous body; the other, that the phenomenon is altogether dependent on an undulation among the particles of a very subtile elastic fluid diffused through space,—as sound is dependent on an undulation among air-particles. Now if light be particles darting around, their minuteness must be wonderful, as a taper can fill with them for hours a space of four miles in diameter ; and with the extreme velocity of light, if its particles possessed at all the property of matter called inertia, their momentum should be very remarkable,—it being found, however, that even a large sunbeam collected by a burning-glass, and thrown upon the scale of a most delicate balance, has not the slightest effect upon the equilibrium. Such, and many other facts to be treated of in subsequent parts of this work, lead to the opinion that there is an undulation of an elastic fluid concerned in producing the phenomenon of light.

" *Becoming less intense as it spreads.*" (See the Analysis, page 162.)

Any emanation from a central point, in spreading through wider space, becomes proportionally thinner or less intense. Thus, if a taper be placed in the centre of a box, each side of which is a foot square, all the light must fall on the sides of the box, and will have a certain intensity there :—if the taper be then placed in a box with sides of two feet square, there will be only the same quantity of light, but that will be spread over four times the surface (a square of two feet is made up of four squares of one foot), and will therefore on any part of that surface be only one-fourth part as strong or intense as in the first box :—and so for any other size of box or space, the intensity diminishing as the square of the distance increases.

Hence four times as much light and heat fall upon a foot of this earth's surface as upon a foot of the surface of the planet Mars, which is twice as distant from the sun :—as four times as much light and heat fall on a man who is at one yard from the fire, as on another who is at two yards.

" *Falling on other bodies makes them visible.*" (Read the Analysis, page 162.)

If the window-shutter of an apartment be perfectly closed, an eye there turns upon an absolute blank : it perceives nothing. If a ray of the sun be then admitted, and made to fall upon any object, that object becomes bright, and affects the eye as if it were itself luminous. It returns a part of the light which falls upon it, and it is visible in all directions, proving that it scatters the received light all around. This scattered light, again, falling on other objects, and reflected from and among them until absorbed, like echo repeated many times and lost between perpendicular rocks, makes all of them also visible, although in a less degree, and the whole apartment is said to be lighted. If the sun's ray be made to fall upon a thing which from its nature reflects much of the light, as a sheet of white paper, the apartment will be well lighted :—if, on the contrary, it be received on black velvet, which returns hardly any light, the apartment will remain dark ;—and, again, if received on a polished mirror, which returns nearly the whole light, but in one direction only, and therefore throws it upon some other single object, the effect will be according to the nature of that object, and nearly as if the ray had fallen directly upon it.

Now all bodies on earth, and very remarkably the mass of atmosphere surrounding the earth, retain and diffuse among themselves for a time the light received directly from the sun, and by so doing, maintain that milder radiance so agreeable to the sight, which renders objects visible when the sun's direct ray does not fall upon them. But for this fact, indeed, all bodies shadowed from the sun, whether by intervening clouds or by any other more opaque masses on earth, would be perfectly black or dark; that is, totally invisible. And without an atmosphere, the sun would appear a red-hot orb in a black sky. On lofty summits, where half the atmosphere is below the level, the direct rays of the sun are painfully intense, and the sky is of darkest blue.

A shadow is the name given to the comparative darkness of places or objects, prevented by intervening obstacles from receiving the direct rays of some luminous body shining on the things around. The apparent darkness of a shadow, however, is not proportioned to its real darkness, but to the intensity of the surrounding lights. A landscape may be very bright, even when the sun is veiled by clouds, and then little or no shadow is perceived; but as soon as the clouds pass away, deep shadows are cast behind every projecting object. Yet the objects and places then appearing so dark, are in reality more illuminated than before the shadow existed, for they are receiving, and again scattering new light from all the more intensely illuminated objects around them. A finger held between a candle and the wall, casts a shadow of a certain intensity; if another candle be then placed in the same line from the wall, the shadow will appear doubly dark, although in fact more light will be reaching the eye from it than before. If the candles be separated laterally, so as to produce two shadows of the finger, but which coincide or overlap in one part, that part will be of double darkness, as compared with the remainders. The most accurate mode of comparing lights is to place them at different distances from a screen or wall, so as to make them at the same time throw equally dark shadows; and then, according to the law of decreasing intensity explained above, to calculate the intensities of the sources of light by the difference of their distances from the wall. The eye judges very easily of the equal intensity of compared shadows.

The real darkness of a shadow depends on the number and nature of the light-reflecting objects around it. Thus, shadows are less remarkable opposite to any white surface, as that of a recently painted wall. The reason why the moon when eclipsed, that is, as will be afterwards explained, when passing through the shadow cast by the earth on the side away from the sun, is almost quite invisible, is, that there are no similar bodies bearing laterally on the moon to share their light with it. And the reason why our nights on earth are darker than the shadows behind a house or rock in the sunshine of day, is merely that there are not other earths near us to reflect light into the great night-shadow of the earth, as there are other houses and rocks to illumine the day-shadow of these. The moon is the only light-reflecting body which the earth has near it; and we perceive how much less dark the night-shadow is when the moon is so placed as to bear upon it. The eclipsed moon, again, is invisible, because facing the shadowed part of the earth; but when the moon is in the situation called new moon, the bright crescent, or part directly illuminated by the sun, is always seen to be surrounding the shaded part, as if holding the old moon in its arms :—that is, the shaded side of the moon is then, in a degree, visible to us, because facing the enlightened side of the earth.

Many persons have doubted whether the light of the moon could be altogether reflected light of the sun; the moon appearing to them more luminous than any opaque body on earth merely exposed to the sun's rays. Their error has arisen from their contrasting the moon while returning direct sunshine with the shadows of night on the earth around them. But could they then see on a hill near them, a white tower or other object scattering light as when receiving the rays of a meridian sun, that object would appear to them to be on fire, and therefore much brighter than the moon. The moon, when above the horizon in the day-time, is perfectly visible on earth, and is then throwing towards the earth as much light as during the night; but the day-moon does not appear more luminous than any small white cloud, and although visible every day except near the change, many persons have passed their lives without ever observing it. The full moon gives to the earth only about a one-hundred-thousandth part as much light as the sun.

" *Light moves with great velocity.*" (See the Analysis, page 162.)

The extraordinary precision with which the astronomical skill of modern times enables men to foretell the times of remarkable appearances or changes among the heavenly bodies, has served for the detection of the fact, that light is not an instantaneous communication between distant objects and the eye, as was formerly believed, but a messenger which requires time to travel :—and the rate of travelling has been ascertained in the same way.

The eclipses of the satellites or moons of the planet Jupiter had been carefully observed for some time, and a rule was obtained which foretold the instants in all future time when the satellites were to glide into the shadow of the planet, and disappear, or again to emerge into view. Now it was found, that these appearances took place $16\frac{1}{2}$ minutes sooner when Jupiter was near the earth, or on the same side of the sun with the earth, than when it was on the other side ; that is to say, more distant from the earth by one diameter of the earth's orbit, and at all intermediate stations the difference diminished from the $16\frac{1}{2}$ minutes, in exact proportion to the less distance from the earth. This proves then that light takes

$16\frac{1}{2}$ minutes to travel across the earth's orbit, and $8\frac{1}{4}$ minutes for half that distance, or to come down to us from the sun.

The velocity of light, ascertained in this way, is such, that in one second of time, *viz.* during a single vibration of a common clock pendulum, it would go from London to Edinburgh and back 200 times, and the distance between these is 400 miles.   This velocity is so surprising that the philosophic Dr. Hooke, when it was first asserted that light was thus progressive, said he could more easily believe the passage to be absolutely instantaneous, even for any distance, than that there should be a progressive movement so inconceivably swift.   The truth, however, is now put quite beyond a doubt by many collateral facts bearing upon it.

As regards all phenomena upon earth, they may be regarded as happening at the very instant when the eye perceives them ; the difference of time being too small to be appreciated :—for, as shewn in the preceding paragraph, if our sight could reach from London to Edinburgh, we should perceive a phenomenon there in the four-hundredth part of a second after its occurrence.

It is hence usual and not sensibly incorrect, when we are measuring the velocity of sound, as when a cannon is fired, by observing the time between the flash and the report, to suppose that the event takes place at the very moment when it is perceived by the eye.

In using a telegraph, no sensible time is lost on account of light requiring time to travel.   A message can be sent from London to Portsmouth in a minute and a-half; and at the same rate a communication might pass to Rome in about half an hour, to Constantinople in forty minutes, to Calcutta in a few hours, and so on. A telegraph is any object which can be made to assume different forms or appearances at the will of an attendant, and so that the changes may be distinguished at a distance. A pole with moveable arms is the common construction, each position standing for a letter, or cypher, or word, or sentence, as may be agreed upon. Telegraphic signals between ships at sea are generally made by a few flags, the meanings of each being varied by the mast on which it is hoisted, and by its combination with others.

" *Light proceeds in straight lines*," &c. (Read the Analysis, page 162.)

Our very notion of a straight line is taken from the direction in which light moves :—but we can verify a line so obtained by other means, as by stretching a cord between the two extremes, or by suspending a weight by a cord, and making a moveable solid measure to correspond with this, which measure may be used in any other case.

We can see through a straight tube, but not through a crooked one. The vista through a long straight tunnel is striking as an illustration of this fact, and of the diminution in the apparent size of objects as they are more distant. If a person enter one end of the canal-tunnel two miles long, cut through the chalk-hills near Rochester to join the Thames and Medway rivers, the opening at the distant end is seen as a minute luminous speck, having the form of the general arch, and appearing in the centre of the shade to an eye placed in the centre ; and a person who has advanced half way through the tunnel, may see the luminous speck, at each end, then appearing a little larger than in the former case.

In taking aim with gun or arrow, we are merely trying to make the projectile go to the desired objects nearly by the path along which the light comes from the object to the eye.

A carpenter looks along the edge of a plank, &c. to see whether it be straight.

Because light moves in straight lines, if a number of similar objects be placed in a row from the eye, the nearest one hides the others. In a wood or a city, a person sees only the trees or houses that are near.

Some ignorant people believe that a squinting person can see round a corner, as they believe that a crooked gun can shoot round a corner.

All astronomical and trigonometrical observations are made on the faith of this property of light, the observer holding that any object is situated from him in the direction in which the light comes to him from it. When the mariner, after watching for hours in cloudy weather, has caught a glimpse of the sun or a star through his sextant-glass, he has ascertained his place among the trackless waves, and boldly advances through the midst of hidden dangers. And the beam from the light-house looking from the rocky height

over the sea, would be useless if the light from
it came not in a straight line.

" *Leaving shadows where it cannot fall.*"   (See the
Analysis, page 162.)

The form of shadows proves that light moves in
straight lines, for the outline of the shadow is
always correctly that of the object as seen from
the luminous body. If the light bent round the
body, this could not be.

The shadow of a face on the wall is a correct
profile.

As a wheel presented edgeways to the eye appears only like a broad line, but becomes oval or
round as it is more turned, so a wheel presented
edgeways to the sun or other light, casts a linear
shadow on the wall behind it, the shadow becoming oval or round as the position is changed.

A globe, a cylinder, a cone, and a flat circle,
will all throw the same form of shadow if held
with their axes pointing to the luminous body, and
therefore by the shadow only, these objects could
not be distinguished.

The figure of a rabbit cut in pasteboard, will
throw the same shadow on the wall as the animal
itself; and, again, that shadow may be perfectly
imitated by a certain position of the two hands
joined, as is known to those who find pleasure
in witnessing the surprise and delight of infancy
made to behold such a shadow mimicking the actions of life.

A man under the vertical sun stands upon his
little round shadow ; but as the sun declines in the

afternoon, the shadow juts out on the opposite side, and at last may extend over a whole field.

A distant cloud which appears to the eye of an observer only as a line in the sky, may be shadowing a whole region; for clouds generally form in level strata, and when viewed at a distance are seen edgeways.

The velocity of the wind may be ascertained by marking the time which the shadow of a cloud takes to pass over a plain or other space of known dimension.

All the heavenly bodies of the solar system cast a shadow beyond them on the side opposite to the sun, as is seen when any body previously visible passes where that shadow is. The satellites or moons of Jupiter, when they suddenly disappear to our glasses, have generally only plunged into the shadow of the planet, and are not hidden behind its body, as many suppose. When our own moon is the subject of that phenomenon so awful in the early ages of the world, an eclipse, she is only passing through the round shadow which the earth casts beyond it.

When the luminous centre is larger than the body which casts the shadow, the shadow will be less than the body. This is true of the shadows of all the planets and of the earth, because they are less than the sun.

On the contrary, if the light-giving surface is smaller than the opaque body, the shadow will be larger than the body. The shadow of a hand held between a candle and the wall is gigantic; and a small pasteboard figure of a man placed near a narrow centre of light, throws a shadow as big as a real man. The latter fact has been amusingly illustrated by the art of making phantasmagoric shadows.

When the surface which receives a shadow is not directly exposed to the light, the shadow may be much larger than the object, even although the sun himself be throwing the light;—as is seen when a slightly projecting roof shadows from the high sun of summer noon the whole front of a house, or as is proved by the long evening shadows of all countries.

" *Light passes readily through some bodies—which are therefore called transparent; but when it enters or leaves their surfaces obliquely, its course is bent.*" (Read the Analysis, page 162.)

It may well excite the surprise of inquirers that light, of which the constituent particles are so inconceivably minute, should still be able to dart readily and in every direction through great masses of solid matter, but such is the truth. Thick plates of solid glass, blocks of rock crystal, mountains of ice, &c., are instantly pervaded by the beam of the sun.

What it is in the constitution of one mass as compared with another, which fits the former to transmit light, and the latter to obstruct it, we cannot clearly explain, but we perceive that the arrangement of the particles has more influence than their peculiar nature. Nothing is more opaque than thick masses of the metals, but nothing is more transparent than equally thick masses of the same metals in solution, nor than the glasses of which a metal forms a large proportion. The thousand salts formed by the union of the metals or earths with the diluted acids, are all transparent, when, in cooling from the fluid to the solid state, their particles have been allowed to arrange themselves according to the laws of their mutual attraction, that is to say, to form crystals ; but the same substances in other states, as when reduced to powder, are opaque. Even the pure metals themselves, when reduced to leaves of great thinness, are transparent, as may be perceived by looking at a lamp through fine gold leaf. It is to be remarked, however, that even the most transparent bodies intercept a considerable part of the light which enters them : a depth of seven feet of pure water intercepts about one-half, so that the bottom of the sea is very dark. And of the sun's light, when passing obliquely through the atmosphere towards the earth, only a small part arrives.

Light having once entered a transparent mass of uniform nature, passes forward in it as straightly as in a vacuum ; but at the surface, whether on entering or leaving it, if the passage be oblique, and if the mass be of a different density from the transparent medium around it, a very curious and most important phenemenon occurs, viz. the light suffers a degree of bending from its antecedent direction, or a refraction, proportioned to the obliquity.

But for this fact, which to many persons might at first appear a subject of regret, as preventing

the distinct vision of objects through all transparent media, light could have been of little
utility to man. There could have been neither
lenses as now, nor any optical instruments, as
telescopes and microscopes, of which lenses form
a part ; nor even the eye itself.

Light falling from the air directly or perpendicularly upon a surface of water, glass, or any
such transparent body, passes through without
suffering the least bending ;— a ray, for instance,
shot from *a* to the point *c*,
in the surface of a piece
of glass *g h*, would reach
directly across to *b ;* but
if the ray fell obliquely,
as from *d* to *c*, then, instead of continuing in its
first direction, and going
on to *i* and *k*, it would at
the moment of its entrance

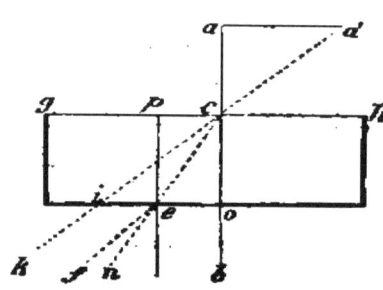

be bent downwards into the path *c e*, nearer to a
line *c o*, called the perpendicular to the surface
at the point of entrance,—and then moving
straightly while in the substance of the glass, it
would, when it passed out again at *e*, in the opposite surface, be bent just as much as at first,
but in the contrary direction, or away from a similar perpendicular at that surface, *viz*. into the
line *e f* instead of *e n*. A ray therefore passing
obliquely through a transparent body of parallel
surfaces, has its course shifted a little to one side
of the original course, but still proceeds in the
same direction, or in a line parallel to the first—

as here shewn in the line _e f_, parallel and near to the line _i k._

The degree of bending or refraction of light in traversing a transparent surface is ascertained by comparing the obliquity of its approach to the surface with the obliquity of its departure after passing ; and for this purpose a line is supposed to be drawn perpendicularly through the surface at the point where the ray passes (as _a b_ in the above figure drawn through _c_, where the ray _d c_ passes), and the relative positions of the ray to this line on both sides of the surface, are easily ascertained. Thus the line _a d_, drawn from any point of such perpendicular to the ray before passing, is a measure of the original obliquity or angular distance of the ray, and is called the _sine of the angle of incidence_, and the other line _o e_ drawn from a corresponding point of the perpendicular to the ray after the passing, is a measure of the obliquity after refraction, and is called the _sine of the angle of refraction :_—by comparing these two lines in any case, the problem is solved.

When light passes obliquely from air into water, the refraction or bending produced is such, that the line _a d_ measuring the obliquity before refraction, is always longer than the line _o e_ measuring it after refraction, by nearly one-third of the latter, and the refractive power of water is therefore signified by the index $1\frac{1}{3}$ or 1.33 ; as in like manner the greater refractive power of common glass has the index $1\frac{1}{2}$, that of diamond the index $2\frac{1}{2}$, and so on. And it is important to remark, that for the same substance, whatever relation holds between the obliquity of a ray and the refraction in any one case, the same holds for all cases. If, for instance, where the obliquity, as measured by its *sine*, is 40, and the refraction is half, or 20, then in the same substance an obliquity of 10 will occasion a refraction of 5, and an obliquity of 4 will occasion a refraction of 2 ; and so on.

As a general rule, the refractive power of transparent substances or media is proportioned to their densities. It increases, for instance, through the list of air, water, salt, glass, &c. But Newton, while engaged in his experiments upon the subject, observed that inflammable bodies had greater refractive power than others, and he then hazarded the conjecture, almost of inspired sagacity, and which chemistry has since so remarkably verified, that diamond and water contained inflammable ingredients. We now know that diamond is merely crystallized carbon, and that water consists altogether of hydrogen or inflammable air and oxygen. Diamond has nearly the greatest light-bending power of any known substances, and hence comes in part its brilliancy as a jewel.

No good explanation has been given of the singular fact of refraction ; but to facilitate the conception and remembrance of it, we say that it happens as if it were owing to an attraction between the light and the refracting body or medium. The light approaching from d to c, for instance, may be supposed to be attracted by the solid body below it, so as at the surface to be bent into the direction c e ; and, again, on leaving the body to be still equally attracted and bent back, so as to take the direction *e f,* instead of *e n.*

The following are familiar examples of this bending of light in passing from one medium to another.

If an empty basin or other vessel *b c f,* be placed in the sun's light, so that the rays falling within it may reach low on the side, as to *d,* but not to

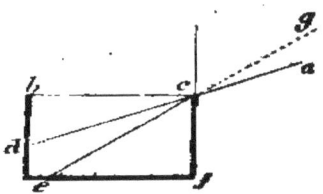

the bottom, then, on filling the vessel with water, the sun will be found to be shining on the bottom or down to *e,* as well as on the side. The reason of this phenomenon is, that water being a denser medium than air, the light, on entering it at *c,* is bent towards the perpendicular at the point of incidence, or *c f,* and so reaches the bottom. Again, if a coin or medal were laid on the bottom of such a vessel at *e,* it would not, while the vessel were empty, be seen by an eye at *a,* but would be visible there immediately on the vessel being filled with water ;—because then, the light leaving it in the direction *e c,* towards the edge of the vessel, would at *c,* on passing from the water into air, be bent away from the perpendicular *c h,* and instead of going to *g* would reach the eye at *a.* The coin moreover would appear to the eye to be in the direction *c d,* instead of in the true direction *c e :* for the eye not being able to discover that the light had been bent in its course, would judge the object to be in the line by which the light came from it.

It is thus because objects at the bottom of water, when viewed obliquely, do not appear so low as they really are, that a person examining a river or pond, or any clear water, from its bank, naturally judges its depth to be less than the truth. Many a young life has been sacrificed to this error. A person looking from a boat directly down upon objects at the bottom of water, sees them in their true places and at their true distances, but if he view them more and more obliquely, the appearance is more and more deceiving, until at last it represents them as at less than half of their true depth.

The ship in which the author sailed once in the China sea, before danger was apprehended, had entered by a narrow passage into a horse-shoe enclosure of coral rocks. When the alarm was given, the predicament had become truly terrific. On every side, in water most singularly transparent, as the wave swelled, the rocks appeared to be almost at the surface of the water, and the anchor, which in the first moments had been let go to limit the danger, appeared to be lifted with them. It was judged that if the ship, then drawing 24 feet, or the depth of a two-storied house, had moved but a little way in almost any direction, she must have met her certain destruction. On sending boats around to sound and to search, the place of entrance was again discovered, and was safely traversed a second time as an outlet from that terrible prison.

On account of this bending of light from objects under water, there is more difficulty in hitting them with a bullet or spear. The aim by a person not directly over a fish must be made at a point apparently below it, otherwise the weapon will miss it by flying too high. The spear is sometimes used in this country for killing salmon, but is a common weapon among the islanders of the Atlantic and Pacific Oceans for killing the albacore ; the use of it, like that of the fly-hook in England, affording, to the fishermen, good sport as well as profit. The author once with much interest witnessed at St. Helena this employment of the spear. A small fish previously half-killed, that it might not try to escape, was every minute or two thrown upon the water as a bait, in the sight of perhaps a hundred great albacores, greedily waiting for it at one side below, and knowing the danger to which they exposed themselves by darting across to seize it. Some albacore bold enough, soon made at the mouthful, apparently with the speed of lightning, but yet with speed which did not save him, for every now and then the thrown spear met the adventurer, and held him writhing there in a cloud of his death-blood. After a victim so destroyed, the scene of action was changed.

The bending of light when passing obliquely from water, is also the reason of the following facts. A straight rod or stick, of which a portion is immersed in water, appears crooked or broken at the surface of the water, the portion immersed seeming to be bent upwards. That part of a ship or boat visible under water, appears much flatter or shallower than it really is. A deep-bodied fish seen near the surface of water, appears almost a flat fish. A round body there appears oval. A gold fish in a vase may appear as two fishes, being seen as well by light bent through the upper surface of the water, as by straight rays passing through the side of the glass. To see bodies under water, in their true places and of their true proportions, the eye must view them through a tube, of which the distant end, closed with plate-glass, is held in the water.

As light is bent on entering from air into water, glass, or other substance denser than air, so is it also bent on coming from void space into the ocean of our atmosphere. Hence none of the heavenly bodies, except when directly over our heads, are seen by us in their true situations. They all appear a little higher than they really

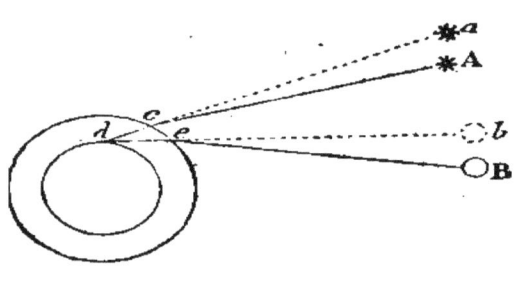

are, as when to a spectator at *d*, supposed on the surface of the earth, a star really at A appears to be at *a*, because its ray on reaching the atmosphere at *c* is bent downwards. In astronomical books there is always introduced a table of refraction, as it is called, shewing what correction must be made on this account for different apparent altitudes. This effect of our atmosphere so bends the rays of the sun that we see him in the morning before he is really above the horizon, and we see him in the evening after he is really below it, *viz.* the ray coming horizontally from *e* to *d*, appears to come from *b*, although in truth it comes from the lower situation B, and is bent into the level line only at *e*. Our atmosphere thus, by the bending of light as well as by itself becoming luminous, lengthens at dawn and twilight the duration of the lovely day. As the atmosphere is denser near the surface of the earth than higher up, the light is more and more bent as it descends, and hence describes a course which is a little curved, and therefore unlike the course of light in water.

Certain states of the atmosphere depending upon its humidity, warmth, &c., change very considerably its ordinary refractive power ; hence in one state of it, a certain hill or island may appear low and scarcely rising above the intervening heights or ocean, while in another state, the same object shall be seen towering above : and from a certain station, a city in a neighbouring valley may be either entirely visible, or it may shew only the tops of its steeples, as if the bed on which it rested had sunk deeper into the earth. In days of ignorance and superstition, such appearances have sometimes excited a strange interest.

A beautiful phenomenon is observable in a day of warm sunshine, owing to the bending of light in passing through media of different densities. Black or dark-coloured substances, by absorbing much light and heat from the sun's rays, and warming the air in contact with them, until it dilates and rises in the surrounding air, as oil rises in water, cause the light, from more distant objects, reaching the eye through the rarefied medium, to be bent a little; and owing to the heated air rising irregularly under the influence of the wind and other causes, these objects acquire the appearance of having a tremulous or a dancing motion. In a warm clear day, the whole landscape at last appears to be thus dancing.

The same phenomenon is to be observed at any time, by looking at an object beyond the top of a chimney from which hot air is rising. An illicit distillery was once discovered by the exciseman happening thus to look across a hole used as the chimney, although charcoal was the fuel, and there was no vestige of smoke.

This bending of light by the varying states of the atmosphere makes precaution necessary in making very nice geometrical observations :—as in measuring base-lines for the construction of maps or charts.

As it is the obliquity between the passing ray and the surface, which in any case of refraction determines the degree of bending, a body seen through a medium of irregular surface appears distorted according to the nature of that surface. It is because the two surfaces of common window-glass are not perfect planes, and not perfectly parallel to each other, as in the case of plate-glass, that objects seen through the former appear generally more or less out of shape : and hence comes the elegance and beauty of plate-glass windows : and hence the singular distortion of things viewed through that swelling or lump of glass which remains where the glass-blower's instrument was attached, and which appears at the centre of certain very coarse panes.

The refraction or bending of light is interestingly exemplified in the effect of the glass called a prism, *viz.* a wedge or three-sided rod of glass, such as that of which the end is here represented

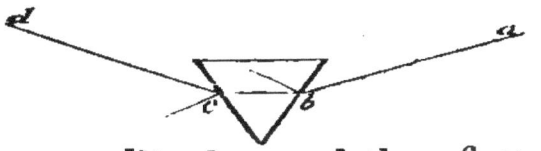

at *b c*. A ray from *a* falling on the surface at *b* is bent *towards* the internal perpendicular, and therefore reaches *c*, but on escaping again at *c*, it is bent *away* from the external perpendicular, and thus with its original deviation doubled, goes on to *d*.

The law of light's bending according to the obliquity with which it traverses the surfaces of a transparent body, is well elucidated by the effect of what is called a multiplying glass; that is to

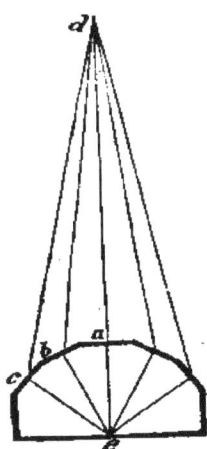

say, a piece of glass like *a b c e*, having many distinct faces cut upon it at angles with each other. If a small object, a coloured bead for instance, be placed at *d*, an eye at *e* will see as many beads as there are distinct surfaces or faces on the glass; for first, the ray *d a* passing perpendicularly, and therefore straight through, will form an image as if no glass intervened, then, the rays from *d* to the surface *b* will be bent by the oblique surface, and will shew the object as if it were in the direction *e b*; then the light falling on the still more oblique surface *c*, will be still more bent, and will reach the eye in the direction *c e* exhibiting a similar object also in that direction—and so of all the other surfaces. If the relative places of the eye and object be changed, the result will still be the same. A plate of glass roughened, or cut into cross fur-rows, becomes a very good screen or window-blind, by so disturbing the passage of light through it that objects beyond it are not distin-guishable.

" *And a beam of white light thus made to bend, is resolved into beams of the various primary co-lours ; which beams however, on being again blended, become white light as before.*" (Read the Analysis, page 162.)

The most extraordinary fact connected with the bending of light is, that a pure ray of white light from the sun admitted into a darkened room by a hole in the window-shutter, and made to bend by passing through transparent surfaces which it meets very obliquely (as the ray *a*, admit-ed and made to bend by passing through the prism of glass *b c*, to fall upon the wall at *d*), instead of bending all together and appearing still as the same white ray, is divided into several rays, which falling on the white wall, are seen to be of different most vivid colours. The original white ray is said thus to be analyzed or divided into elements.

This solar spectrum, as it is called, formed upon the wall, consists when the light is admitted by a narrow horizontal slit, of four coloured patches corresponding to the slit, and appearing in the order, from the bottom, of red, green, blue, and violet. If the slit be then made a little wider, the patches at their edges overlap each other, and as a painter would say, produce by the mixture of their elementary colours various new tints. Then the spectrum consists of the seven colours commonly enumerated and seen in the rainbow, *viz.* red, orange, yellow, green, blue, indigo, and violet. Had red, yellow, blue, and violet been the four colours obtained in the first experiment, the occurrence of the others, *viz.* of the orange, from the mixing edges of the red and yellow— of the green, from the mixture of the yellow and blue,—and of the indigo, from the mixture of blue and violet, would have been anticipated. But the true facts of the case not being such, proves that they are not yet well understood. When Newton first made known the pheno- menon of the many-coloured spectrum, and the extraordinary conclusions to which it led, he ex- cited universal astonishment, for the common idea of purity the most unmixed, was that of white light. In farther corroboration of the notion of the compound nature of light he mentioned, that if the colours which appear on the spectrum be painted separately round the rim of a wheel, and the wheel be then turned rapidly, the individual colours cease to be distinguished, and a white band only appears where they are whirling: also, that if the rays of the spectrum produced by a prism be again gathered together by a lens, they re-produce white light. The red is the kind of

light which is least bent in refraction, and the
violet that which is most bent. It was at one
time said, as an explanation, that the differently
coloured particles in light had different degrees
of gravity or inertia, and were therefore not
all equally bent. It is farther remarkable, with
respect to the solar spectrum, that much
the heat in the ray is still less refracted than even
the red light, for a thermometer held below the red
light rises higher than in any part of the visible
spectrum ;—and that there is an influence more
refrangible than even the violet rays, producing
powerful chemical and magnetical effects. The
different spots of colour are not all of the same
size, and there is a difference in this respect according to the refracting substance.

All transparent substances in bending light produce more or less of the separation of colour;
but it is important to remark, that the quality of
merely bending a beam, or of *refraction*, and
that of dividing it into coloured beams, or of
dispersion, are distinct qualities, and by no means
proportioned to each other in the same substances.
Newton, from not discovering this, despaired that
a perfect telescope of refraction could ever be
made : he supposed that the bent light would
always become coloured, and so render the objects indistinct. We now know, however, that
by combining two or more media, we may obtain
bending of light without dispersion,—thus, by
opposing a glass which bends five degrees and
disperses one degree, to another glass which
bends three degrees and disperses one, the oppo-

sing dispersions will just counterbalance or neu-
tralize each other, while the two degrees of excess of bending will remain to be applied to use.

The diversified colours of the substances around
us depend merely upon the fitness of these, from
texture or other cause, to reflect or transmit certain modifications of common light, and the colour
is not a part or property of the body itself. We
shall soon find that the vivid colours of the rainbow are merely the white light of the sun, reflected to us after being bent and modified by the
colourless drops of falling rain; and that the
with appearance of rubies and emeralds,
ee in the cut-glass lustre, is a phenomenon of the same kind :—and that by scratching
the surface of a piece of metal so as to have a given
number of lines in a given space, we can cause
the same substance to appear of any colour we
please.

" *Transparent bodies, as glass, may be made of such
form as to cause all the rays of light which pass
through from any one point, to bend so as
to meet again in another point beyond them,—the
body itself, from the required form generally resembling that of a flat bean or lentil, being then
called a* LENS." (Read the Analysis, page 162.)

The innumerable rays of light, of which five
only are here represented, issuing from any point
as *c*, towards any surface in the situation *a b*,
are said to form a cone or pencil of diverging light.
Now it is evident that to make all such rays converge or meet again in one place, as *f*, beyond a

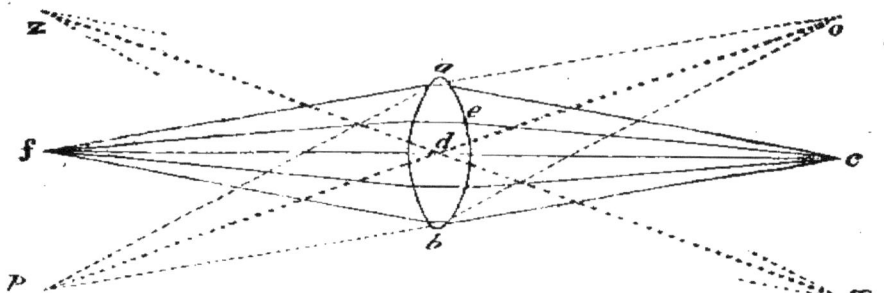

transparent body placed at *a b*, it would be necessary, while the middle ray or axis of the pencil
c d did not bend at all, for the others to be bent
more and more, in proportion as they fell upon the
body farther and farther from the centre *d*. Recollecting then the law of refraction, that light entering from air through the surface of any denser
medium, as glass, is bent there towards the perpendicular at the internal surface, in proportion
to the obliquity of incidence, and on leaving the
opposite surface, is correspondingly bent away
from its external perpendicular (see the case of
the prism at p. 188), we see that if a piece of
glass were placed at *a b*, of such form that the
rays falling upon it from *c* should meet and leave
its surfaces with greater and greater obliquity in
some regular proportion, as the points of incidence
were more distant from the centre *d*, the purpose
would be obtained. And we have the pleasure of
knowing that a glass, of which the surface is
ground—which it easily may be—to have a regular convexity or bulging, as if it were a portion
cut off from the surface of a globe, can be
shewn to answer very correctly the required condition. Such a glass, similarly ground on both
sides, is here represented edgeways, between *a*

and *b*, where the ray *c d* falling on its middle, or perpendicularly, and similarly leaving it, is seen going straight through to *f*; but the ray *c e* meeting the surface with a certain degree of obliquity, is bent down a little, first on entering the surface at *e*, and then as much more on leaving the opposite surface with equal obliquity, and so arrives at *f*; then the ray *c a*, for corresponding reasons, is still more bent, and equally arrives at *f*;—and the case would be similar of any other rays that might be examined. The point *f* is usually called a *focus* (meaning a fire-place), because when the light of the sun is thus gathered, the heat concentrated with it is powerful enough to make combustibles inflame.—We have here to remark farther, that in accordance both with calculation and experiment, the direction in which a pencil of rays falls upon a lens does not affect the result of the convergence to a focus, only the focus is always in the direction of the central ray of the pencil or beam ; it will be at *p*, for instance, for light issuing from *o*, and at *z* for light issuing from *x*.

The lens represented at *a b* above, or at fig. 1, in the annexed dragram, having both sides convex is called *a double convex lens.* A glass convex only on one side, and plane or flat on the other,

Fig.1   2   3   4   5   6 as shewn at (fig 2), would as effectually gather the rays, but with

half the power, and the point of meeting or focus

would be therefore so much the more distant.
Such a glass is called a *plano-convex lens*.   Then
the gathering or converging power of any glass,
whether doubly or singly convex, is in proportion
to the degree of its convexity or bulging of surfaces, for the less it bulges, the more nearly does
it approach to a plane glass, and the more it
bulges, the more obliquely will the rays at any
distance from the centre fall upon its surface, and
the sooner therefore, in consequence of their being
more bent, will they all meet the axis-ray ;—hence
fig. 1 would converge much more quickly than
fig. 3, which represents nearly a common spectacle glass ; and a very minute globe is the form
most powerfully converging of all.   The surfaces
of fig. 1, are portions of a comparatively small
globe ; those of fig. 3 are comparatively smaller
portions, but of a globe much larger.   Concave
lenses as—fig. 4, a double concave, and fig. 5, a
plano-concave lens, in obedience to the same law
of refraction, spread rays, or bend them away
from the axis of the pencil, in the same degree
that similarly convex lenses gather them.   A concave lens therefore, receiving the converging
pencil of rays from a convex lens, might restore
them to their former direction.   Very useful purposes, as will be afterwards explained, are served
in optics, by certain combinations of differently
formed lenses.   A lens may be convex on one side
and concave on the other, as at fig. 6, a meniscus
lens (so called because it resembles the crescent
moon), and its effect will be according to the form
which predominates.

A person recollecting the case of the " multiplying glass," described a few pages back, might say,—but is not a convex lens merely a multiplying glass of a much greater number of faces, and why then, instead of one image, does it not make thousands ? The answer is, that the multiplying glass, by every face, bends *a set* of rays, capable of forming a distinct image : but the lens has no surface large enough to bend more than a single ray, and it concentrates all the single rays into one place, to form there one image of great vividness and beauty.

" *And when the light proceeding from every point of an object placed before a lens is collected in corresponding points behind it, a perfect image of the object is there produced. When the image is received upon a suitable white surface, in a dark place, the arrangement is called, according to minor circumstances, the* CAMERA OBSCURA, SOLAR MICROSCOPE, *or* MAGIC LANTERN." (Read the Analysis, page 162.)

Words are wanting to express the admirable consequences to man of the curious property of a lens, that it can bring together to a focal point all the rays of light which traverse it from any one point of an object placed before it. The following instance will lead to the consideration

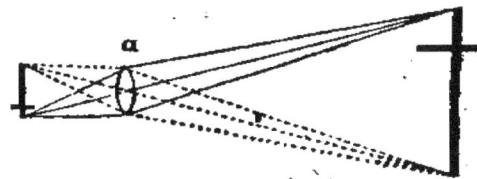

and understanding of others. If a lens, as *a*, be placed so as to fill up an opening made in the window shutter of a darkened room, then from any object before that opening, as the cross here represented, all the light which each point of the cross emits towards the lens will be concentrated or gathered together in a corresponding focal point behind the lens or within the room, and if a sheet of paper be held there at the distance of the focal points, a beautiful image of the object will be seen upon the paper.

In these few words, we have described the interesting contrivance called the *camera obscura* or *dark chamber ;* and when a glass is chosen of proper size and focal distance, and a screen or the wall of the chamber is properly prepared to receive the light, the most enchanting portraiture is instantly produced of the whole scene which the window commands. With what rapture does the school-boy first view this lovely picture drawn by nature's own pencil, and with colours borrowed directly from the sun's bright ray—with what rapture, as his eyes search over it, does he recognize his playmates there, and perhaps the river in which he bathes, and where he sails his boat, and the wood in whose solitudes he loves to wander, and the mountain heights which he climbs to meet the fresh breeze, and at a distance from the world, to allow play to the workings of his young fancy, beginning to shoot far into time and space. The great peculiarity of such a picture is, that it does not, like others, pourtray still-nature, but every thing with appropriate motion or changes : there the playmates are all in action ; the leafy trees may wave in the wind, the clouds may sail along, the sun may rise or may set, and even the lightning's gleam may dart across :—or, again, commenced enterprizes may be brought to a close, the traveller may climb the distant hill and disappear, the fisherman may draw his net and store his gains, the well-contested race may be won or lost. A Malayan chief in the island of Sumatra, was so surprised and pleased by a small portable camera obscura which the author had among his apparatus, that he seemed disposed to give for it almost any thing he possessed.

It appears in the last diagram that the image formed beyond a lens by the gathered light, is in a contrary position to the object itself,—that is, inverted,—because, the light from the top of the object darts through the opening or glass in a descending direction, and that from the bottom rises to the opening, and in the same direction passes beyond it. It is usual therefore in a camera obscura to place a small mirror immediately behind the lens, so as to throw all the light which enters, downwards to a whitened table, upon which the picture may be conveniently contemplated.

The camera obscura often gives very useful assistance to young painters, by enabling them to trace correctly the outlines of the objects placed before it, and also to study effects of light, shade, and colour, more profitably than they at first can, by looking at the objects themselves. The laws of perspective are most intelligibly illustrated in this most true picture.

An effect approaching in a degree to that of the complete camera obscura now described, is produced by merely making a small hole in the shutter of a dark room, and letting the light which enters by it fall on any white surface beyond. The whole landscape is then dimly pourtrayed upon the surface. If a cross be held before the opening as in the last figure, it is evident that from every point of the cross light will enter by the opening, and will fall on corresponding parts of a sheet of paper held behind,—but as the light from each point spreads as it departs, becoming a pencil or cone of light instead of a ray, it will fall on a surface of the paper at least as large as the opening, and thus the light from adjoining spots will mix at the edges, and will render the images misty and indistinct, somewhat like those on the back of tapestry. If the opening be very small, the picture will be well defined, but very feebly illuminated ; and if the opening be of considerable size, the mixing of the pencils will be so great as to leave no particular object distinguishable. In the latter case, and supposing the opening to be larger than the pupils of a thousand eyes, if a lens be introduced, it will converge every pencil of light to an exact point, and the picture will instantly be rendered perfectly clear.

The distance from a lens at which an image is formed or the rays of the light meet, depends, first, upon the refractive or bending power of the lens, and therefore on the nature of its substance, and the form of the lens ; and, secondly, upon the direction of the rays of light when they reach the lens, *viz.* whether they are divergent, parallel, or convergent. We have already explained that glass refracts about twice as much as water, and that diamond refracts about twice as much as glass ; and we have considered the effect of different degrees of convexity in lenses—arising equally whether the lens be of water inclosed between glasses like watch-glasses, or of solid glass, or of rock-crystal, or of diamond itself. We now proceed to consider the joint effect of the refractive power, and of the direction of the incident rays.

Rays falling from *a* on a comparatively flat or weak lens at L, might meet only at *d*, or even

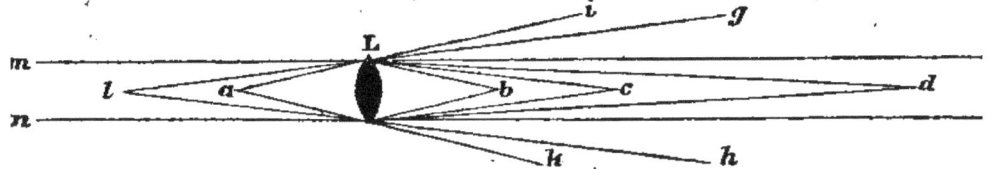

further off; while, with a stronger or more convex lens, they might meet at *c* or at *b :* a lens weaker still might only destroy the divergence of the rays, without being able to give them any convergence, or to bend them enough to bring them to a point at all,—and then they would proceed all parallel to each other, as seen at *e* and *f :* —and if the lens were yet weaker, it might only destroy a part of the divergence, causing the rays from *a* to go to *g* and *h*, after passing through, instead of to, *i* and *k*, in their original direction.

In an analogous manner, light coming to the lens in the contrary directions from *b c d*, &c., might, according to the strength of the lens, be all made to come to a focus at *a* or at *l*, or in some more distant point; or the rays might become parallel, as *m* and *n*, and therefore never come to a focus, or they might remain divergent.

It may be observed in the figure above, that the farther an object is from the lens, the less divergent are the rays darting from it towards the lens; or the more nearly do they approach to being parallel. From *b* there is much divergence, from *c* less, from *d* less still, and rays from a great distance, as those cut off at *e f* appear quite parallel. If the distance of the radiant point be very great, they really are so nearly parallel that a very nice test is required to detect the non-accordance. Rays, for instance, coming to the earth from the sun, do not diverge the millionth of an inch in a thousand miles. Hence where we wish to make experiments with parallel rays, we take those of the sun.

Any two points so situated on the opposite sides of a lens, as that when either becomes the radiant point of light, the other is the focus of such light, are called *conjugate foci*. An object and its image formed by a lens must always be in *conjugate foci*, and when the one is nearer the lens, the other will be in a certain proportion more distant.

What is called the principal focus of a lens, and by the distance of which from the glass we compare or classify lenses among themselves, is the point at which the sun's rays are made to meet; and thus, by holding the glass in the sun, and noting at what distance behind it the little luminous spot or image of the sun is formed, we can at once ascertain the focus of a glass—as at a for the rays e and f.

It is remarkable that the bending power of the common glass used for lenses should be such, that the focus of a double lens of glass is just where the centre of the sphere would be, of which the surface of the lens is a portion. This gives us another fact with which to associate the recollection that the focus is near as the convexity of the lens is greater, that is to say, as the surface is a portion of a smaller sphere. And such being the law, it may be proved by calculation as well as by the fact, that if a candle be held from a lens at twice the principal focal distance, suppose at c for a lens with the focus at a, the image of the candle will be formed at l just as far on the other side. Thus then, by trying with a lens until the image of a candle is formed at the same distance from it as the object is, we have a second mode of ascertaining the focal distance of a lens. Other kinds of glass and other substances refract with different power ; but the facts now stated should be retained in the memory as standards of comparison.

Because the focal point of light passing through a lens is at the same distance from the centre of the lens, in whatever direction the light passes through, a surface placed to receive the image of any object should really be concave, that is to say, all parts of it should be at the same distance from the centre of the lens, otherwise the image will be more perfect either at its middle than towards its edges, or *vice versâ*—but it is not found necessary to attend to this in common practice.

The size of an image formed behind a lens is always proportioned to its distance from the lens, and the image is as much larger or smaller than the object as it is farther from or nearer to the lens than the object. This will be evident from

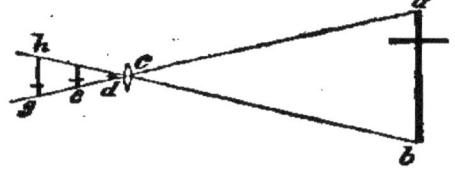

considering the annexed figure. *c* represents a lens, which, according to its power, will form an image of the cross *a b* in some situation, as at *d, e, g,* &c. Now wherever the image is formed, and by whatever lens, one end of it must be in contact with the line *a g,* and the other end with the line *b h*; and as these lines cross each other at *c,* and widen regularly afterwards, a line joining them (and the image is really such a line), must always be shorter the nearer it is to *c,* that is to say, shorter in proportion to the converging power of the lens.

Many persons may not have reflected, that the little luminous circle called the focus of a burning glass, is really but the image or picture of the sun formed by that glass or lens. The intensity of the heat and of the light is of course in proportion as the image is smaller than the glass which forms it, and the nearer that the image is formed to the lens, or the more powerfully convergent that the lens is, the smaller will the image be. Mr. Parker's famous burning lens, which cost £700, and is now the property of the Emperor of China, was three feet in diameter, and the diameter of the sun's image formed by it was one inch : it concentrated therefore about 1,300 times. To render the effect still more powerful, a smaller lens was placed behind the larger, further reducing the size of the image to one-sixth. Very surprising effects were produced by this lens, in the melting of metals, inflaming of combustibles, &c. The size of burning lenses, until lately, was limited by the difficulty of obtaining the great pieces of glass required to form them : but they are now built up of many pieces suitably united together. Some large lenses have been made of water, that is, of water enclosed between meniscus glasses, like watch-glasses. A common goblet of water, or a vase holding gold-fishes, has acted as a burning glass in some cases, where, in consequence of its being left in a sunny window near the curtains, a house has been burned.

And the nearer that an object is brought to a lens, the more distant, and therefore the larger, will its image be : for, as the rays towards a lens diverge in proportion to the nearness of the object, and therefore with the same power of lens, must meet farther behind,—as seen in the figure at page 201, then the axis of the rays, as the lines $c\,a$ and $c\,b$ in the last figure, will have separated far before the rays meet, and will have made the image proportionally larger. If we suppose little d in the same diagram to be the object, its image would be $a\,b$. The sun is exactly as much larger than his image formed by a burning glass, as he is more distant from it than the image ; and if we had a canvas of sufficient size hung up in distant space, a very bright object of a quarter of an inch diameter might be made to form an image as broad as the sun.

From all these considerations, we see that, in a camera obscura, the screen should be from the lens, at the distance of its principal focus for distant objects, and a little farther than this for near objects. Accordingly the lens is generally fixed in a sliding piece, which allows the distance from the screen to be adjusted to circumstances. If the representation is wished to be large, the lens must be of a long focus ; if to be small, the lens must be of a short focus. Again, when by the reversed use of the lens, a small object as d is to be magnified to such a size as $a\ b$, then the object must be placed a little beyond the focus of the glass ; for if placed nearer, the pencils of rays from it would never be gathered to focal points, and no image would be formed at any distance.

When, as alluded to in the last sentence, a small object is placed very near a lens, and the image of it is thrown upon the wall of a dark room, perhaps a hundred times farther from the lens than the object is, the image is a greatly magnified representation of the object, *viz.* it is a hundred times longer and a hundred times broader, and therefore has ten thousand times as much surface as the object ; but if in this experiment the object be illuminated only in an ordinary degree, the light from it is so scattered as not to suffice for distinct vision. Hence, to attain fully in this manner the purpose of a microscope, a very strong light, concentrated by a suitable mirror or glass, must be directed upon the object.

When the light of the sun is used in such a case, the complete apparatus is called the *solar micro-scope*, and serves beautifuily to display the struc-ture of many minute objects. When artificial light is used, as of a lamp, the apparatus is called the *lucernal microscope* or *magic lantern*.

A good solar microscope becomes one of the most interesting presents which science has made to man, for aiding him in his researches into the secrets of nature. With the late improvements in the construction of lenses, by which the disper-sion of light, or the rainbow-fringe, is prevented (as will be explained under the head of " Teles-copes "), objects may be magnified two or three hundred thousand times, and still be so luminous as to be beautifully distinct :—thus a cheese-mite will appear of the dimensions of a little pig, and creatures altogether invisible to the naked eye, or perceived by it only as minute white points, are discovered to be animated beings, having the per-fect proportions, and often the great beauty of larger animals, and endowed with similar appetites, passions, and apparent ingenuity, but with an ac-tivity far surpassing that met with in the more bulky creation. A judicious selection of objects for the solar microscope is calculated exceedingly to surprise the mind on first attending to them, and to fill it with high conceptions of the infinity of God's creation. With the common microscope only one person at a time can feast his wonder ; but with the solar, a whole roomful of company may at once contemplate the same objects and witness the same actions, and thus have their admiration increased by the consciousness of a sympathy.

The magic lantern, we have said, consists of a powerful lens, with objects, highly illuminated by lamp-light, placed so near it that their images are formed far off, and are therefore proportionally larger. For the magic lantern the objects are generally paintings on glass made with transparent colours ; and the glass is formed to slide through a slit or passage behind the lens. The lens itself,—or what may be called half of it, for there are often two lenses joined to give greater power, is moveable with the tube which is seen projecting from the lantern, so that its distance from the object may be varied, and thus a corresponding approach to or receding from the screen may be allowed, which will produce an increase or lessening of the magnitude of the visible picture on the wall.

Some public lecturers on astronomy and branches of natural history prefer having the drawings and paintings required for the elucidation of their subjects, made in miniature upon glass, to be magnified afterwards to the degree desired, and shewn upon any part of the lecture-room by the magic lantern.

A thick fog or mist at night will sometimes reflect the images of a magic lantern so as to make them distinctly visible ; and there are several cases on record, where persons, wickedly ingenious in this way, have terrified ignorant individuals almost to death, by throwing spectres from a concealed lantern. Some years ago a sentinel in St. James' Park was thus persuaded that he had seen supernatural beings near him among the trees.

A very charming illusion is produced by a magic lantern manœuvred on one side of a thin screen, let down like the curtain of a theatre, while the spectators, not aware of the existence of the screen, are sitting on the other side. The image may be first thrown upon the screen with the lantern very near, and then it will be small, and exceedingly bright if desired, because the light is much concentrated. By the exhibitor then gradually receding from the screen, and at the same time adjusting the distance of the lens from the picture, the image—which may be that of a genius flying in the air, becomes larger, and to the spectators who do not even know that there is a screen there, it will appear to be soaring and approaching—until at last the expanded wings and limbs may seem hovering almost over their heads. An endless variety of most ingenious and beautiful exhibitions of this kind have been made under the name of the *phantasmagoria,* or *raising of spectres.*

" *And the* EYE *itself is, in fact, but a small camera obscura,—of which the pupil is the round opening or window before the lens.* (Read the Analysis, page 162.)

The Eye.—And who could at first believe that in describing the camera obscura, as we have now done, we had in reality been describing that most interesting of the objects of creation, the living eye itself, the great inlet of man's knowledge, that which may be called the visible dwelling of the soul, or at least the window of that dwelling —that from which all the fire of passion darts, through which the languor of exhaustion is perceived, in which life and thought seem concentrated! Yet the eye is nothing but a simple camera obscura, formed of the parts described above as essential to the camera obscura :—but in its simplicity it is so perfect, so unspeakably perfect, that the searchers after tangible evidences of the existence of an all-wise and good Creator, have declared their willingness to be limited to it alone in the midst of millions, as their one triumphant proof. We shall now describe it and its actions. Keeping present to us the idea of the camera obscura, as already treated of, the use of the various parts of the eye will be declared by merely enumerating them. This paragraph should be perused while the reader has the opportunity of observing either his own eye reflected in a glass, or the eye of some companion near him.

The *human eye* then is a globular chamber of the size of a large walnut, formed externally by a very tough membrane called, from its hardness, the *sclerotic coat*, in the front of which there is one round opening or window, named, because of its horny texture, the *cornea.* The chamber is lined with a finer membrane or web—the *choroid,* which to insure the internal darkness of the place, is covered with a black paint, the *pigmentum nigrum.* This lining at the edge of the round window is bordered by a folded drapery—the *ciliary processes,* hidden from without by being behind the curious contractile window curtain the *iris,* through the central opening of which, or *pupil,* the light enters. Immediately behind the pupil is suspended by attachments among the ciliary processes, the *crystalline lens,* a double convex most transparent body of considerable hardness, which so influences the light passing through it from external objects, as to form most perfect images of these objects in the way already described, on the back wall of the eye, over which the optic nerve, then called the *retina,* is spread as a second lining. The eye is maintained in its globular condition by a watery liquid, which distends its external coverings, and which in the compartment before the lens, or *the anterior chamber of the eye,* being perfectly limpid, is called the *aqueous humour,* and in the remainder or larger *posterior chamber,* being enclosed in a transparent spongy structure, so as to acquire somewhat of the appearance of melted glass, is called the *vitreous humour.*

The annexed figure represents an eye of the common dimensions, supposed to be cut through

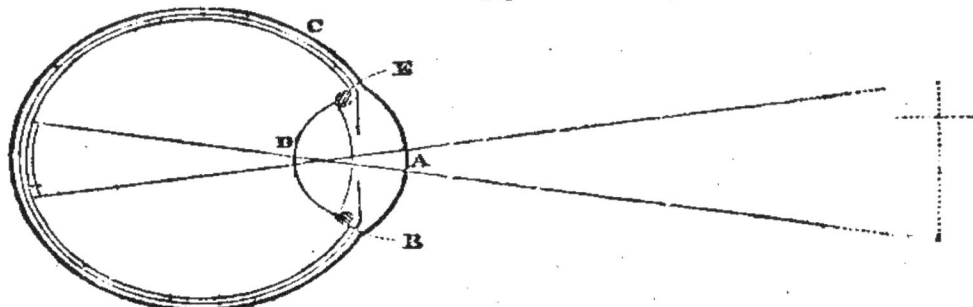

the middle, from above downwards. C is the outer or *sclerotic* coat, known popularly, where most exposed in front, as the *white of the eye.* A is the

transparent cornea joined to the edge of the round
opening of the sclerotic : it is more bulging than the
sclerotic, or forms a portion of a smaller sphere than
the general eye-ball, so that while it may be truly
called a *bow-window*, it, or rather the convex surface
of its contained water, is also a powerful lens for
acting on the pencils of entering light.   At B,
and similarly all round the edge of the cornea, is
attached the window curtain or *iris*, shewn here
edgeways, immersed in the aqueous humour, and
hanging inwards from above and below towards
its central opening or *pupil*, through which the
rays of light are passing to the lens.   The iris has
in its structure two sets of fibres, the circular and
the radiating, which cross and act in opposition
to each other.   When the circular fibres contract,
the pupil is lessened, when the radiating contract,
it is enlarged ; and the changes happen according
to the intensity of light and the state of sensibility
of the retina,—as may at any time be proved by
closing the eyelids for a moment to make the
pupil dilate, and then opening them towards a
strong light, to make it contract.   Behind the
pupil is seen the *lens* D with its circumference attached to the *ciliary processes* E : it is more convex
behind than before.   The disease of the eye called
cataract (from a Greek word implying *obstruction*),
is the circumstance of the lens becoming opaque,
and the cure is to extract the lens entirely, or to
depress it to the bottom of the eye, and then to
substitute for it externally a powerful artificial
lens or spectacle-glass.   The three lines forming
here the boundary of the eye stand for its three

coats as they have been called, the strong *sclerotic,* and the double lining of the *choroid* and *retina.* The figure of a cross is represented upon the retina as formed by the light entering from the cross without, which cross has to appear here small and near, although supposed to be large and distant. The image of the cross is inverted, as explained for the camera obscura : but we shall learn below that the perception of an object may be equally distinct in whatever position the image be on the retina. It has been explained above, that a lens can form a perfect image of considerable extent only on a concave surface,—and the retina is such a surface. The present diagram further explains what is meant by the *anterior* and *posterior chambers* of the eye, *viz.* the compartments which are before and behind the crystalline lens D.

The nature of the eye as a camera obscura is beautifully exhibited by taking the eye of a recently killed bullock, and after carefully cutting away or thinning the outer coat of it behind, by going with it to a dark place and directing the pupil towards any brightly illuminated objects ; then, through the semi-transparent retina left at the back of the eye may be seen a minute but perfect picture of all such objects—a picture, therefore, formed on the back of the little apartment or camera obscura, by the agency of the convex cornea and lens in front.

Understanding from all this, that when a man is engaged in what is called looking at an object, his mind is in truth only taking cognizance of the picture or impression made on his retina, it excites admiration in us to think of the exquisite delicacy of texture and of sensibility which the retina must possess, that there may be the perfect perception which really occurs of even the separate parts of the minute images there formed. A whole printed sheet of newspaper, for instance, may be represented on the retina on less surface than that of a finger-nail, and yet not only shall every word and letter be separately perceivable, but even any imperfection of a single letter. Or, more wonderful still, when at night an eye is turned up to the blue vault of heaven, there is pourtrayed on the little concave of the retina the boundless concave of the sky, with every object in its just proportions. There a moon in beautiful miniature may be sailing among her white edged clouds, and surrounded by a thousand twinkling stars, so that to an animalcule supposed to be within and near the pupil, the retina might appear another starry firmament with all its glory. If the images in the human eye be thus minute, what must they be in the little eye of a canary-bird, or of another animal smaller still! How wonderful are the works of nature!

Because the images formed on the retina are always inverted as respects the position of the objects producing them—just as happens in a simple camera obscura, persons have wondered that things should appear upright, or in their true situations. The explanation is not difficult. It is known that a man with wry neck judges as correctly of the position of the objects around him as any other person—never deeming them, for instance, inclined or crooked, because their images are inclined as regards the natural perpendicular of his retina; and that a bed-ridden person obliged to keep the head upon the pillow, soon acquires the faculty of the person with wry neck : and that an affected girl inclining her head while trying her attitudes, from much practice judges of the manœuvres of a beau as conveniently in that way as in any other; and that boys who at play bend down to look backwards through their legs, although a little puzzled at first, because the usual position of the images on the retina is reversed, soon see as well in that way as in any other. It appears therefore, that while the mind studies the form, colour, &c. of external objects in their images projected on the retina, it judges of their position by the direction in which the light comes from them towards the eye—no more deeming an object to be placed low because its image may be low in the eye, than a man in a room into which a sun-beam enters by a hole in the window-shutter, deems the sun low because its image is on the floor. A candle carried past a key-hole, throws its light through to the opposite wall, so as to cause the luminous spot there to move in a direction the opposite of that in which the candle is carried ; but a child is very young who has not learned to judge at once in such a case, of the true motion of the candle by the opposite apparent motion of the image. A boatman, who, being accustomed to his oar, can direct its point against any object with great certainty, has long ceased to reflect, that to move the point of the oar in some one direction, his hand must move in the contrary direction. Now the seeing things upright, by images which are inverted, is a phenomenon akin to those which we have reviewed.

Another question somewhat allied to the last is, why, as we have two eyes, and there is an image of any object placed before them formed in each —why the object does not appear to us to be double. In answer to this, again, we need only to state the simple facts of the case. In the two eyes there are corresponding points, such that when a similar impression is made on both, the sensation or vision is single : but if the least disturbance of the position occur, the vision becomes double. And the eyes are so wonderfully associated, that from earliest infancy they constantly move in perfect unison. By slightly pressing a finger on the ball of either eye, so as to prevent its following the motion of the other, there is immediately produced the double vision ; and tumours about the eye often have the same effect. Persons who squint have always double vision : but they acquire the power of attending to the sensation in one eye at a time. Animals which have the eyes placed on opposite sides of the head, so that the two can never be directed to the same point, must have in a more remarkable degree the faculty of thus attending to one eye at a time.

The corresponding points in the two eyes are equidistant and in similar directions from the centres of the retinæ, called the points of distinct vision, at which centres the imaginary lines named the axes of the eyes terminate; and it is worthy of remark that these points, in being both to the right or both to the left of the centres, must be one of them on the inside of the centre, as regards the nose, and the other on the outside. When the two eyes are directed to any object their axes meet at it, and the centres of the two retinæ are opposite to it, and all the other points of the eyes have perfect mutual correspondence as regards that object, giving the sensation of single vision; but the images formed at the same time, of an object nearer to or farther from the eye than the first supposed, cannot fall on corresponding points, for an object nearer than where the axes meet would have its images on the outsides of the eyes, and an object more distant would have its images on the insides of the eyes, and in either case the vision would be double. Thus if a person hold the two fore-fingers in a line from his eyes, so that one may be more distant than the other, by then looking at the nearest, the more distant will appear double, and by looking at the more distant, the nearer will appear double.

The reason of the term " point of distinct vision," applied to the centre of the retina, is discovered at once by looking at a printed page, and observing that only the one letter to which the axis of the eye is directed, is distinctly seen; so that although the whole page be depicted on the retina at once, the eye, in reading, directs its centre successively to every part.

On examining a dead eye, the point of distinct

vision is distinguishable from the retina around by being more transparent. Now it might have been expected that this point would have been the spot where the optic nerve enters the eye : but in truth the optic nerve enters considerably nearer to the nose than the centre of the retina ; and very singularly, where it enters, the part is altogether blind or insensible. Had the two optic nerves then entered at *corresponding* points of the retina (in the sense explained above), there would have appeared a black spot on every object opposite to the insensible points ; but as the case really stands, the part of any object from which the light passes to the insensible part of one eye must be opposite to a sensible part of the other. The existence of the insensible or blind spot, where the nerve of the eye enters, is discoverable by placing in a row three objects—wafers, for instance—on a table, with intervals of about two inches between them, and then looking with one eye from a distance of about eight inches at the wafer which is towards the nose ;—the middle wafer will then be invisible, although the eye sees that on each side of it ; and if the eye be moved still farther away, the middle wafer will come into view, and the external will disappear. Or, again, the fact may be proved by shutting one eye and looking with the other at the nail of a finger held before it, while another finger is gradually moved away laterally : the point of the moving finger when at a certain distance from the other will disappear, but will be seen again when moved away still a little further.

It appearing then, from the explanations now given, that there cannot be perfect sight unless where a perfect image is formed on the retina; and the truth having been formerly explained, that images behind any lens will be at different distances from it, according to the various distances of the objects in front, that is to say, according as the pencils of light which fall upon it have more or less of divergence in them, it follows, that the eye in being able, as it is, to see distinctly objects at any distance beyond about five inches, possesses a power of altering the relation of its parts to accommodate itself to the circumstances. We do not yet perfectly know whether it does this by lengthening or changing the form of the ball through the action of the surrounding muscles, or by changing the place or the form of the lens, but that one or more of these events occurs there can be no doubt.

Among the eyes of the myriads of human creatures, however, it happens that all do not originally possess these powers exactly in the requisite degree, and that many lose them, as life advances, from a natural or usual decay.

Persons are called *short-sighted* whose eyes from too great convexity of the cornea or lens, have so strong a bending or converging power, that the

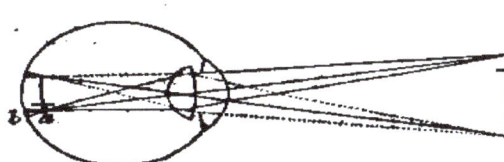

rays of light entering them are brought to a focus before reaching the retina—at *a*, for instance, instead of at *b* ; so that the

rays, by spreading again beyond the focus, produce on the retina that sort of indistinct image
which is seen in a camera obscura of which the
screen is too distant from the lens. This defect
of sight obliges the individual when using the
naked eye to hold objects very near it, that the
consequent greater divergence of the rays may be
proportioned to the unusual refractive power of the
eye ;— or the person may find a remedy in placing
concave lenses between the object and the eyes,
which lenses, by rendering light from objects at a
usual distance more divergent (as explained at
page 195), cause the perfect images in the eye to
be formed farther from the lens, and thereby on
the retina itself. Without concave spectacles—
as the lenses are called when fixed together in a
frame—persons with the defect now under consideration cannot see distinctly any object that is
distant, and from which the rays, because coming
nearly parallel, are quickly gathered to a focus.
This defect often diminishes with years, and the
person who in youth needed spectacles, in old age
sees well without them.

There is an opposite defect of deficient converging power in the eye, dependent on a too
great flatness of the cornea or lens, which defect
is much more common than the last-mentioned,
for the great majority of persons after middle age
sooner or later begin to experience it. In this
case the rays of light are not yet collected into a
focus when they reach the retina ; they would only
meet at b, for instance, instead of at c, and hence
the image is indistinct, in the same manner as in

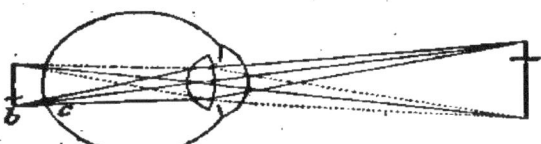

a camera obscura of which the screen is held too near the lens. Persons suffering this defect cannot, when using the naked eye, see distinctly any object very near to it, because the gathering or converging power of the eye cannot conquer the great divergence of rays coming from a near point ; and hence such persons always remove objects under examination to a considerable distance, often to that of arm's length, so as to receive from them only the rays nearly parallel. These persons, in contradistinction to the last described, are called *long-sighted* persons. Their defect is remedied by the common convex spectacles, which do part of the converging work, so to express ourselves, before the light enters the eye, leaving undone only that which the eye can easily accomplish. As this ailment, like the last, is met with in all degrees, it becomes requisite to choose spectacles accordingly : certain curvatures or strengths have been numbered as naturally belonging to different ages or periods of life, but each person should choose under the direction of an experienced judge, until that strength is found which enables him to read, without any straining of the eye, at the common distance of from twelve to eighteen inches.——We cannot apply the mind to this part of our subject without feeling admiration at what science has accomplished for man in guarding and improving his sight. Now that in civilized society the most common employments and enjoyments of life are such as to require visual power capable of distinguishing minute objects,—letters for instance, to deprive old men of their spectacles, would be to condemn many of them to useless inactivity and a listless blank of mind for the remainder of their lives.

An eye much accustomed to examine near and minute objects, often loses something of its pliancy, and becomes defective when tried at distant things, as the watchmaker's eye, the engraver's, &c. On the other hand, the old seaman, whose eye has so often and uninterruptedly been bent on the distant horizon, straining to catch the view of an expected sail, or of land, has a power of discovering distant things which is wonderful; but he often experiences some deficiency in regard to near things.

A man who uses his eyes under water sees very indistinctly, because the difference of density between the two transparent media—water and the eye, from the first of which, in the case supposed, the light passes to the second, is not so great as between air and the eye, the bending or refraction of the light is consequently not so great. A man, to see well under water, therefore, requires to aid the usual power of his eyes, by strong convex spectacles. It is for the reason now explained, that the lens of a fish's-eye is extremely convex : indeed it is almost round, as is every day seen in the white round bead which issues from the eye of a boiled fish—that little globe being the crystalline lens of the fish coagulated or hardened like white of egg during the cooking.

There are many important considerations connected with the sensibility of the retina, which regard rather the laws of life than of light, but we must here glance at a few of them.

Any impression of light made upon the retina lasts for about the sixth of a second. Hence when the burning end of a stick is made to describe any line or curve, its path becomes a line of light; and if it revolve in a circle six times in a second, that circle will appear to the eye a complete circle of fire. The polished end of an elastic wire fixed by its other end in a block of wood, being made to vibrate, similarly forms a line or curve of light. A harp-string while vibrating as it sounds, appears like a flat riband. Lightning or other meteor darting across the sky, although in fact but a moving luminous point, is generally thought of as a long line of light: the term forked-lightning has reference to this prejudice. The same remark applies in a degree to a sky-rocket in its rapid ascent. Two or more colours painted separately on the rim of a wheel which is made to turn rapidly, appear to the eye to be as completely united as if they were really mixed: —it has been already explained how patches of the various colours of the rainbow mixed in this way form white light. If on one side of a card a little bird be painted, and on a corresponding part of the other side a cage, then on making the card turn rapidly by twisting between the fingers two threads fixed to its opposte edges, the little bird will appear to be imprisoned in the cage: or, again, if a pensive Juliet sitting in her bower oc-

cupy one side of the card, and a longing Romeo
the other, by the magic turn of the threads the
passionate lovers may instantly be brought together.   Dr. Paris displayed taste and an amiable
ingenuity in designing this toy with great variety
of subjects.

A certain intensity of light is necessary
distinct vision, but the degree varies with the
previous state of the organ.   A person passing
from the bright day into a shaded room, for a
time may fancy himself in total darkness; and to
persons sitting in the room and become accustomed to the less light so as to see well with it, he
will appear to be almost blind.   The dawn of
morning after the darkness of night appears much
brighter than an equal degree of light in the
evening.   When, as the night falls, our lamps or
candles are first introduced, the glare is often for a
time offensive : and the same feeling is still stronger
on opening, in the morning, bed-room window-
shutters or close-drawn curtains.   After the repose of night, the sensibility of the eye is such
that the globules of blood in the capillary vessels
of the retina produce the impression on it of little
globes of light crossing among each other as the
tortuous vessels do.   To a prisoner after long
confinement in a dark dungeon, the light of the
sun is almost insupportable.   And a dungeon,
which to common eyes is utterly dark, still to its
long-held inmate has ceased to be so.   There are
various instances in the records of the barbarous
ages, of prisoners confined for years in utter darkness, who at last could see and make companions

of the mice which frequented their cells. The darkness of a total eclipse after bright sunshine appears much more deep than that of midnight, because of the sudden contrast. The long polar night of months ceases to appear very dark to the polar inhabitants. If an eye be directed for a time to a black wafer laid on a sheet of white paper, and afterwards to another part of the sheet, a portion of the paper of the size of the wafer will appear brilliantly illuminated: for the ordinary degree of light from it appears intense to the part of the eye lately receiving almost none. An eye directed long and intensely upon any minute object—as when a sailor watches a speck in the distant horizon, supposed to be a ship, or when a hunter on the brown heath keeps his eye fixed on some game nearly of the colour of the heath, or when an astronomer gazes long at a little star—has the sensibility of its centre at last exhausted, and ceases to perceive the object ; but on directing the axis of the eye a little to one side of the object, so that an image may be formed only *near* the centre, the object may be again perceived, and the centre in the mean time enjoying repose, will recover its power.

But the most extraordinary fact connected with the sensibility of the retina is, that if part of it be strongly exercised by looking for a time at an object of any bright colour, on then turning the eye away or altogether shutting it, an impression or spectrum will remain of the same form as the object lately contemplated, but of a perfectly different colour. Thus if an eye be directed for

Q

a time to a red wafer laid on white paper, and be then shut or turned to another part of the paper, a beautifully bright green wafer will be seen, and *vice versâ*, a green wafer will produce a red spectrum, an orange wafer will similarly produce a blue spectrum, a yellow one a violet spectrum, &c.; and a cluster of wafers will produce a similar cluster of opposite colours. If the hand be then held over the eyelids to darken the eyes and prevent entirely the approach of light, the spectrum of the bright parts will be luminous surrounded by a dark ground, and when the hand is again removed the contrary will be true. Again, if the eye be in a degree fatigued by looking at the setting sun, or even at a window with a bright sky beyond it, or at any very bright object, on then shutting it, the lately contemplated forms will be perceived, first of one vivid colour, and then of another, until perhaps all the primary colours have passed in review. These extraordinary facts prove that the sensation of light and colour, although excitable by light, is also producible without it. This truth gave occasion to Darwin's ingenious theory, that the sensation of any particular colour, of red for instance, is dependent upon a certain state of contraction of the minute fibres of the retina, as the sensation of a particular tone depends on a certain frequency of vibration of some part of the ear,—and that the fibres, when fatigued in that condition, seek relief when at liberty, by throwing themselves into an opposite state,—as a man whose back is fatigued by bending forward, relieves himself not by merely standing erect, but by bending the spine backwards—which new condition, whether produced by light or by any other cause, gives the sensation of green. He applied his explanation similarly to all other cases of colour. It is remarkable that the colours which thus appear opposite to each other in kind are those which when the solar spectrum produced by a prism, as described a few pages back, is painted round a wheel or circle, are opposite to each other in place.

There are persons who although having distinct perceptions of form and of light and shade, have not the power of distinguishing colours. It is common for such persons to deem pink and peagreen (naturally opposites) the same colour, and therefore not to distinguish difference of colour in a red berry and the leaves around it. A man with this defect, trusting to his own judgment, might, without knowing it, dress himself like a parrot.

" *The mind judges of external objects by the relative size, brightness, colour, &c. of the minute but perfect images of them formed at the back of the eye on the expansion of nerve called the retina ; and the art of painting is successful in proportion as it produces on a larger scale a picture, which when afterwards held before the eye to reproduce itself in miniature upon the retina, may excite the same impression as the original object.*" (Read the Analysis, page 162.)

We now understand how an admirable miniature resemblance of the objects before us is produced upon the retina of the eye, by the light from them refracted in passing through the different parts of the eye; but after all, this is only a picture, and the inquiry remains—which many persons would suppose so simple as to be trifling, but which is in reality most curious and important—how we are thereby enabled to judge of the magnitudes, distances, and other particulars respecting the things examined. Here it will be found, to the surprise of persons first entering upon the subject, that we learn the meaning of a scene or pictorial signs only gradually, as we do of any other system of signs, and that a person no more sees, in the complete sense of the word, that is to say, no more understands any scene or prospect when he first opens his eyes upon it, and has a perfect picture of it on his retina, than he understands or can read a printed page, on first looking into a book before he has learned his letters. Most interesting information has been obtained on this subject, by observing the facts where blindness from birth has been, by surgical operation, suddenly cured in persons arrived at maturity.

If a man were placed from infancy in an apartment fitted up as a camera obscura, and had no means of becoming acquainted with external nature, but by watching the images appearing upon the screen, he could learn almost nothing of the universe around him; but if after a time he were allowed to walk out, and to examine by the touch and by measurement the different objects pourtrayed there, and to ascertain what size, shape, and distance of an object corresponded with a certain magnitude, form, position, and brightness of image, the imagery might at last be to him a very clear indication of such particulars, and through them of nearly all else that he desired to know ; making him in imagination present to the objects around, almost as if he went and examined them with his hands like a blind man, or in any other way. In the same manner, nearly, the soul may be considered as if originally placed in the little camera obscura of the eye, where it has to acquire experience of external nature by commanding the services of the bodily limbs or members. The judging of things by sight, then, is merely the interpreting one set of signs, as judging by sounds or language is interpreting another, and judging by hieroglyphics or any written characters is interpreting a third. The common visual signs on the retina, however, are those most readily learned or understood, from having certain relations in form, &c. to the things signified.

Bodies differ and are distinguished among themselves chiefly by their comparative dimensions, that is, their size and shape ; and to ascertain these and the relative distances, are the great objects which by the eye the mind seeks to accomplish. Now it effects its ends by considering collectively,

1st. *The space and place* occupied by objects in the field of view, measured by what is called the *visual angle.*

2d. *The intensity of light, shade, and colour.*

3d. *The divergence of the rays of light.*

4th. *The convergence of the axes of the eyes.*

We shall treat of these particulars separately in the order now mentioned.

   1st. *The space and place occupied in the field of view, measured by the visual angle.*

The field of view is that open or visible space before the eyes, in which objects are seen; and the term may mean either the small field visible in one position of the eyes, or that which is perceived on directing them all around. As the eye may be turned in every direction, it may be considered as placed in the centre of a hollow sphere, where it sees the several objects around occupying certain situations and certain proportions of the circum-

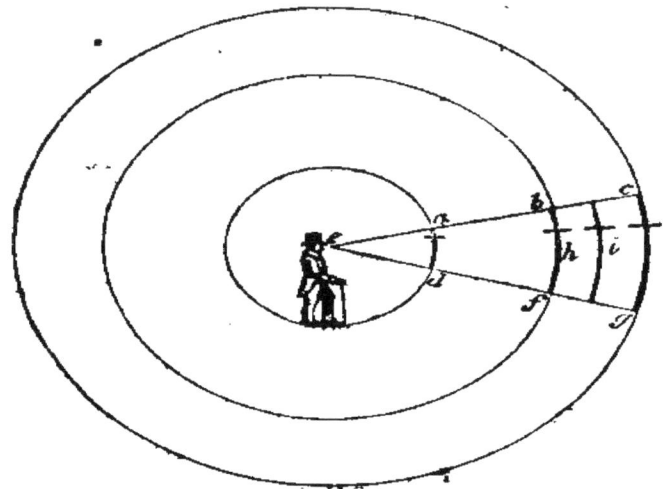

ference: and if a man were really sur-rounded by a large globe or sphere of glass as *a*, through which he might view objects, which sphere had any equal divisions or degrees marked upon it all around, he would be able at once to say exactly what portion of his sphere or field of view was shadowed or occupied by any single object, as the cross here shewn at *i*, and thus to describe very intelligibly, either for his own recollection, or to inform others, its re-lative magnitude and situation as then appearing to him,—just as he might say, on looking at a tree in the garden through a common window (which is a portion of the field of view really divided by the cross bars), whether he saw the whole tree through one pane or through several, and through which pane or panes he saw it. It may be remarked farther, that whether the supposed sphere of glass were large or small, *viz.* were as *a* or as *b* or *c*, the part of its surface apparently occupied by any object beyond or within it, would bear the same proportion to the whole surface. Now as men have found it convenient to consider a circle (and every circle) as divisible into 360 degrees (which are smaller therefore in a small than in a large circle, although in each having the same relation to the whole), the ready mode of comparing the apparent magnitude of objects is to say how many of these degrees of the field of view—supposed a portion of a hollow sphere surrounding a man, each object occupied : and this is really what is meant by the apparent size of an object. And because the most convenient way of measuring a portion of a circle, of which the whole is not seen, is to measure the angle formed at its centre by lines drawn from the extremities of the portion,—as here the angle at *e* formed by the lines *c e* and *g e*, the object is said to occupy a certain number of degrees of the circumference of the circle, or to subtend an angle of the same number of degrees at its centre, and this angle is called the *visual angle*, the subject of our present disquisition.

The visual angle then, in regard to any object, is that included between the lines or rays, as *a u*

and *d i*, which from the extreme points of the
object, as *a d*, cross in the lens of the eye, and
go afterwards to form the extremes of the image
on the retina, and, as formerly explained, the
angle is the same on either side of the lens, *viz.*
towards the object or towards the image.   Now if
all bodies were at the same distance from the eye,
the magnitudes of their images formed on the retina, or in other words of the visual angles subtended by them, would be an exact measure of
their comparative real magnitudes, as is seen in *i u*,

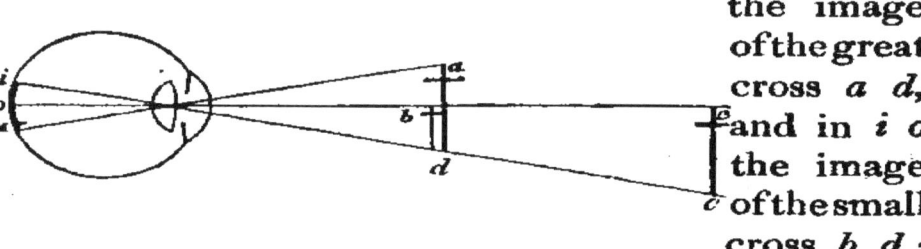

the image
of the great
cross *a d*,
and in *i o*
the image
of the small
cross *b d :*

but it is evident here, that the cross *c e*, which is
twice as large as *b d*, makes, because twice as far
off, an image of only the same size as *b d*, and an
image therefore only half as large as that of a cross
a d equal in size with itself: and the same rule
of proportion holds for all other comparative distances—at a hundred times the distance, an object
appearing only the hundredth part as tall, and so
forth.   To judge therefore by the eye of the true
size of an object, we must know its distance as
well as its apparent size or visual angle.

Many familiar facts receive their explanation from
the law of the visual angle or apparent size being less always in proportion as the distance of
an object is greater.

· A man, (or a cross) at d, standing near the outside of a window, as $b\,c$ (shewn here edgeways), may to a spectator seated within the window at h, subtend the same visual angle, or appear as tall as the window, the light from his head passing through the top of the window, and that from his feet passing

h through the bottom : but if the man then move away from the window, the spectator will be able to see his whole body through a smaller and a smaller extent of the window,—as through half its height or $a\,c$, when he is twice as distant from the eye, or at f, and through the third or $o\,c$, when he shall be three times as distant, or at g, and so forth, for any other distance ; so that soon a small figure of a man cut in paper, if laid upon the glass, would exactly cover the part of it through which the light from him entered to the spectator's eye, and would then, by completely hiding him from view, be an exact measure of his apparent size : and at last a fly passing over the pane might equally hide him, and the fly then would subtend a larger visual angle than he, that is to say, would be forming on the retina a larger image than the man. Thus it often happens in reality, that a person sitting near a window, and intent upon some subject of study or of conversation, mistakes a fly on the glass for a man at a distance ; or, on the contrary, a man for a fly. It is ascertained that the eye, with an ordinary degree of light, can see an object which in the field of view occupies only the sixtieth of a degree (or one minute) in a circle of twelve inches diameter, the eye being supposed in the centre of the circle. This space is about the 100th of any inch measure held six inches from the eye. Now a body smaller than this, at six inches, or any thing, however large, placed so far from the eye as to occupy in the field of view less space than this, is invisible to ordinary sight. At four miles off, a man is thus invisible. A pin-head near will hide a house on a distant hill—nay, will hide even the planet Jupiter, although 1,000 times bigger than this earth.

In accordance with the principle now explained a marine telescope has been constructed, having the field of view divided by fine cross wires, or otherwise, so that the person using it can say at once how much of its field any object occupies. Now when ships are in chase, it is common by this instrument, or some other that will detect a change of apparent size, or of visual angle, to view the fleeing or pursuing ship ; and if the apparent size be observed to increase, it is known that the ships are nearing each other ; if on the contrary it diminish, the chased ship is escaping.

By applying this rule, whenever the exact size of a distant object is known the distance is ascertainable, and, *vice versâ*, where the distance is exactly known the size is determinable :—for it is evident that if a body, as a ship, known to be 100 feet tall, occupy or subtend in the field of vision the 360th part of a whole circle, or one degree, the whole circle must be 360 times 100 feet,

or 36,000 ; and knowing the diameter of such a
circle to be nearly one-third as much, we learn
the distance of the ship, *viz.* half the diameter.
Again, if we know the distance of a ship or other
object to be a mile, and if we then find its visual
angle to be the 1,000th part of a circle, we know
its true size to be the 1,000th part of a circle, of
which the half diameter or radius is one mile. It
is by applying this rule in a manner to be afterwards explained, that we determine the size of the
heavenly bodies.

We now perceive that if the rays of light coming
to the eye through a plate of glass, from objects
seen beyond it, could leave marks in the glass at
the points where they passed, and marks capable
of giving out the same kind of light as caused
them, there would be formed *upon* the glass a representation or picture of the objects formerly
viewed *through* it, and that picture would be so
perfect, that when held before the eye, it would
form on the retina an image or images the same
in almost all respects as the objects themselves
had done ; for from the different points of the
glass, light would dart to the eye in the very same
directions pursued by that originally darted from
the objects. Now the art of painting seeks so to
dispose lights, shades, and colours on any plane
surface, as to produce the sort of representation
of objects here contemplated, while the picture-
frame has to recall the window-frame, or edge of
the plate of glass through which the true scene is
supposed to be viewed. It is remarkable indeed
how perfectly this art now accomplishes its ends ;

and although there are still trifling differences
between the effect upon the eye, of a common
picture, and of the realities,—which peculiarities
we shall consider presently, and how they may
be combated so as to render the illusion quite
perfect,—it is not one of them, as might be supposed from the small extent of the canvas, that
the picture appears to the retina smaller than the
objects themselves. Few people, before studying
this subject, are aware that in a good picture the
size of the figures is always made exactly such,
that at the distance from the eye at which they
are meant to be beheld, they produce on the retina
the very same size of image as would be produced by
the realities seen, under the aspect represented in
the picture. To become sensible of this, let a person
look through a window-pane, with the eye at the distance of eight inches from it, and let him draw with a
sharp point upon the glass previously coated with
gum, the outline of the scene beyond—perhaps a
street or square, he will find, that the outline of a
man seen there at the distance of twenty paces,
appearing perfectly to coincide with the boundaries of the person, and such as, if opaque,
would just hide the person, will be scarcely half an
inch tall, while the figure of a man a few hundred
paces off will appear as a little point too small for
the minuter features to be distinguished, even if
they could be drawn.

Now as a person who reads the description of an
elephant, does not deem the animal larger or
smaller because of the size of letter used in the
printing, or of the size of the accompanying en-

graved representation, whether it be diminutive in the page of a child's nursery-book, or wide spread over the page of a quarto—and as a man, viewing in a picture-gallery miniatures and larger portraits, does not conceive of the originals according to the size of the representations—and as a man who views a picture of a temple, so perfect that it might almost be mistaken for the reality, never dreams, unless his attention be particularly directed to the fact, that the distant pillars of the rows are vastly smaller upon the canvas than the near ones; but in all such cases the mind merely uses the *signs* to help it to conceive of the *things* according to other principles of judging; so in any common case of seeing, the mind takes so little account of the *apparent* size of objects passing instantly from the types to the realities, generally known, that it soon ceases to be aware that the apparent size of the same object ever changes. Most persons would be surprised to be told, for instance, that a man with whom one is shaking hands, appears to the mere eye ten times taller than when he has walked ten paces away, or that a chair at one end of a room appears to a person sitting at the other, only half as large as a chair in the middle of the room; but such are the facts: and they may be immediately proved by holding a common eye-glass or ring at a certain distance from the eye, and then looking through it at any similar objects placed at different distances; then, while of a chair standing near, only a small part will be visible through the ring—of a distant chair, the whole may be seen;—and so of any other case. At five miles distance, Nelson's fleet on the great day of Trafalgar might have been seen through a marriage-ring as the picture-frame. There are occasions, however, where the usual collateral helps to the immediate recognition of objects being wanting, the attention is strongly aroused to the fact of their diminutive appearance produced by distance—for instance, when a man first from the high sea approaches a land, of which the features are in a degree new to him : an Englishman arriving in India has considerable difficulty in believing the little specks which he sees scattered along the shore to be commodious dwellings, and what seem to him only luxuriant herbs or bushes, to be magnificent palm-trees.

For the same reason that a distant body to the mere eye appears diminutive, *viz.* the smallness of the visual angle between the extreme points observed, so does a distant motion to the mere eye appear slow. A carriage dashing past a pedestrian in the street, may surprise him by its speed ; but if viewed at the same time by a spectator from the top of St. Paul's, it seems to be but crawling along the pavement. A ship driven before a tempest, scarcely allows the sailor on board to distinguish the individual masses of the white foam through which she flies ; but if then seen on the distant horizon by a spectator on shore, she is scarcely perceived to change her place. A balloon high in the air, and borne along on the wings of the wind at the rate of seventy or eighty miles an hour, may still for a time leave a spectator on earth doubtful as to the direction in which it is moving. The moon in her orbit wheels round the earth at the astonishing rate of about 2,000 miles an hour, yet, owing to her distance from it, her motion is not there visible to the naked eye, except by comparing her place at considerable intervals. In respect to bodies still more distant than the moon, the truth at present under consideration is still more striking.

Having now explained how the apparent tranverse measure of bodies and of space, in other words the visual angle subtended by them, is affected by their distance from the eye, we proceed to shew how it is affected also by their position.

A globe at a certain distance from the eye, however turned, has the same appearance or bulk in the field of view, and its outline traced upon glass held between it and the eye, is always a circle ; but an egg, although when held in one position it produces a circular outline or image, when held in another, produces an image nearly oval. A wheel when viewed sideways appears a perfect circle, when viewed edgeways it appears a broad straight band or line, and when in any intermediate position it also appears oval. The apparent form then is only a hint to the mind from which, by former experience or other means, it guesses at the true form. If a man had never seen an egg but endways, he never could have known that it was not a perfect sphere.

If any long straight object, as a beam, be placed with one of its ends directly to the eye, that end only can be seen, and according to the case must

appear a square or circle of the diameter of the
beam ; if it then be placed with its side directly
to the eye, its whole length will be seen ; and if
placed in any intermediate position, it will appear
more or less shortened ;——in all cases, its outline on
the retina being similar to that of its shadow on a
wall behind the person. A man has advanced on a
spear pointed directly to his eye without seeing it,
or on a bar of iron carried on the shoulder of a
porter met in the street. A common telescope
held with its end to the eye appears a perfect circle, if then inclined a little, it seems to jut out
on one side, and as the inclination is increased,
it juts out more and more, until it displays its
whole length. A great ship of war whose stern
is towards a spectator, appears a rounded building with its rows of windows like those of a
peaceful habitation ; but as it turns, it gradually
reveals the bristling cannon in their whole length
of fearful batteries. A straight row of a thousand similar objects, as of soldiers in rank, pillars,
trees, &c., may appear to a person at the extremity as only one object of the kind, the nearest
individual completely hiding all the others ; but
if viewed from the side and at a certain distance,
the individuals may be counted.

The appearance now treated of is called *foreshortening,* and is to be noted wherever any surfaces or lines are not placed so as completely to
face the spectator.

Perhaps the most important case of foreshortening which has to be observed is that of an extensive plane surface, along which the eye looks——

for instance, the general surface of the earth or sea, by estimating aright the foreshortening of which, we partly judge of the distance or situation of the objects placed upon it. It is evident that in all such cases the more distant portions of the surface are progressively more foreshortened than the nearer ; for a man standing on a plain as *a b,* and looking down immediately before him with his eye at *c,* sees a portion of the surface almost directly, or without any foreshortening, and an extent of 5 feet, as *a d* (if 5 feet be the height of the eye), will subtend in the eye an angle of 45°, *viz.* the angle *a c d,* or will appear 45° long in his field of view—therefore half of what is subtended by the whole space from his feet to the horizon ; the next 5 feet will subtend an angle of only 18°, *viz., d e f,* the next of 8°, *viz., f c g,* and so on ; and as he carries his view more and more forward, the surface becomes to it more and more oblique, until at last the light coming from the surface seems rather to skim along the level than to rise. This explains why a person having a side view of a row of separate objects, as of men in line, trees, pillars, &c., may see through between the nearest of them, but towards the extremes of the view sees them as if standing in closest possible array, or as if forming a continued surface. The same remark explains why masses of cloud scattered uniformly over the sky, may allow a spectator to see wide intervals of the blue heaven over-head, while all around there is a dense cloudy wall appearing to rest on the horizon.

If a man standing on a hill look down upon a field or plain which is well known to him, and if he see some objects near its side, and some near its middle, and some near its distant border, he knows at once how far they are from him and from each other. Similarly, if viewing the ocean from a lofty cliff, and seeing ships scattered over its face, he may judge correctly of their distance, for he can see only a certain extent of ocean which becomes to him as a known field. The man stationed at the flag-staff on the High Knowl peak of St. Helena, looks down upon a circular field of the Atlantic a hundred miles broad, and he tells the distance of any sail in sight to within a mile or two. Now although the ground plan of a landscape may not be so level as the field or ocean-face now spoken of, there is still an approximation to the true plain, which very considerably assists a spectator's judgment of distances.

Painters are not only careful to foreshorten correctly, according to the proportion explained above, all the objects which they pourtray, but they often avail themselves of the principle to produce most striking effects. For instance, Martin in many of his beautiful designs, by judicious foreshortening, has exhibited miles in extent of gorgeous architecture and of armed men, on a space of canvas that would seem scarcely more than sufficient to receive a few figures :' he has made a single magnificent pillar or accoutred warrior placed in the foreground, become the type which first fills the mind with admiration, and then sends it along the retiring lines of beautiful perspective, where every tip or edge renews the first impression. A man lying on a table or a bed nearly as high as the eye, with his feet towards the spectator, is foreshortened into a roundish heap, of which the soles of the feet hide the greater part. This is the description of the painting which has been called the miraculous entombment of Christ, and it is because an unreflecting spectator moving sideways with the expectation of seeing more of the body, still sees only the soles of the feet, and may suppose the body turned round so as to front him, that the painting has received its appellation. For nearly the same reason the eyes of a common portrait may seem to follow a spectator to whatever part of the room he goes. A rifleman represented as taking aim directly in front of the picture, will seem to have in his power every spectator standing in the room ; for, as in the case of the miraculous entombment, every spectator present will feel as if he alone could see the picture as all see it. To terrify young ladies, a little arch Cupid has ingeniously been represented with his arrow pointed directly at them, and just ready to let it slip from his bended bow :—and, oh, how they are terrified !

As the painter, availing himself of a knowledge of the principles now explained, by which the eye usually judges of size and distance, may produce on his canvas the most charming illusions, so may the tasteful landlord in his ornamental gardens and pleasure grounds, by working his levels into artificial undulation of hill and dale, with magnitude of tree and of edifice to correspond—make the eye of a spectator luxuriate in the contemplation of supposed extensive plains, lofty mountains, distant pagodas, and wide-spread lakes—all within the narrow space of an acre or two ;—thus, in truth, by other means, producing on the retina the same impressions as Claude, Poussin, or Wilson, by their finest pictures.

When any object or mass of objects is foreshortened, by one part being further from the eye than another, that part appears also in a proportion smaller than the other. For example, in a straight row of similar houses, trees, &c., those nearest to the eye will, on a glass held before the eye to receive their images, form the largest images, and there will be a gradual diminution from the largest to the least, so that lines drawn upon the glass along the tops and bottoms of the images would tend to a point, called, for a reason explained below, the *vanishing point*. Thus a person looking from a window upon a long-straight street, must, to see to the chimnies of the nearest house, look through the top of the window, and to see the street-door must look through the bottom ; but the most distant house, both top and bottom, may be concealed from view by a little spot upon the glass at the height of the eye. This remarkable tapering of foreshortened objects may of course be strikingly observed on looking at any correctly made drawing or engraving meant to represent a retiring row of similar objects ;—such drawing being in truth an attempt to realize by art the appearance of the objects as seen through the window.

The art which attempts to trace objects on a plane surface, as they would appear on looking at them through that surface if it were transparent, with their various degrees of apparent diminution on account of distance, and of foreshortening on account of obliquity of position, is called, from the Latin word signifying *to look through*, the *art of perspective*. It consists entirely of the two parts now mentioned; and notwithstanding the terror with which the study of it is clothed in the imaginations of many young painters, by reason of the mathematical difficulties with which it has usually been mixed up, it is in itself exceedingly simple. We hope that a person capable of ordinary attention will, from what we have already said, and from the few additional remarks which we have still to make on the appearances of nature, be able completely to understand the great laws of perpective. Although, without a knowledge of these laws, a quick eye soon enables its possessor to sketch from nature with much truth; and although the two instruments, the *camera obscura* and *camera lucida*, give almost mathematical accuracy to drawings, without requiring other skill in the draughtsman than to trace with ink or pencil the lines which he sees as if on the paper, still the subject is so interesting to all who look either at nature or the works of art, that no intelligent person should neglect it.

Supposing a straight row of similar objects, as of the stone blocks or pillars represented here from *a* to *S*, to be viewed by a person standing

near c, then, because, as already explained, objects to the eye appear smaller in exact proportion to their increased distance from it, the second block, if twice as far off as the first, would appear only half as large ; the third, if three times as far, would be only one-third as large, and so on to any extent, and for any other proportions ; and if the 1,000th or any other nearer or more distant pillar subtended to the eye an angle less than the sixtieth of a degree of the field of view, it would be altogether invisible, even if nothing intervened between it and the eye. Then, where the row ceased to be visible from the minuteness of the parts, or from the fact of the nearer objects concealing the more remote, it might be said to have reached its *vanishing point*.

Now it is very remarkable that in any such case of a straight line or row vanishing from sight, in whatever direction it points, east for instance, although the eye to see the near end of it would have to look about north-east, still the point in the heavens, or in a picture, or transparent plane before the eye, where the line would vanish, would be exactly east from the eye, and not in the slightest degree either to the north or to the south

of the east point, because the pillars happened to
be north or south of the individual; and therefore, if there were two or more rows of pillars
parallel to the first, but considerably apart
from each other, as the lines here *a S, b S, d S,*
&c. still all would vanish or seem to terminate in
the very same point of the field of view. The
reason of this is easily understood. Let us suppose a line drawn directly east from the eye or to
the point *S, viz.* a line directly over the line *c S,*
and that the line of pillars *a S,* also pointing east,
is 20 feet north of the spectator, and the line of
pillars *b S,* running in the same direction, is 20
feet south of him, then evidently for the same
reason as the space between the top and bottom
of the pillars, that is to say their height, becomes
apparently less as their distance from the eye
increases, so will the space between each pillar and
the point corresponding to its place in the visual
ray, or the line along which the eye looks, become
less, and the lines of pillars really 20 feet apart
from the visual ray, will at a certain distance from
the eye, *viz.* where 20 feet is apparently reduced
to a point, appear to join it, and the three lines
will appear to meet in that point, beyond which
they cannot be visible, and which is therefore
called the vanishing point. The conception of
this truth may be facilitated by our supposing a
star or planet to be rising in the eastern point of
the heavens at the moment of observation; then,
if the three parallel lines were continued on to
the planet, and were visible as far, they would
arrive there with the 20 feet of interval between

them just as they left the earth ; but as any planet, although many thousand miles in diameter, owing to its distance from the earth, appears only a point, much more would two lines only 20 feet apart be there undistinguishable in place by human sight. And what is true of a space of 20 feet between parallel lines, is equally true, as regards human vision, of a space of hundreds or of thousands of miles : as a general rule therefore, it holds, that all lines in reality parallel to each other in perspective tend to and finish in the same vanishing point, *viz.* the situation of the line in which the eye looks when directed parallel to any one of those real lines. And this is true not only of lines in the same level or horizontal plane, *viz.* such as might be along the surface of the sea, but also of lines that are vertical or one above another, as those running along the tops and bottoms of the pillars here, or along the roofs and windows of the houses, and indeed of all lines in whatever situation, provided they are parallel to the visual ray. When it is ascertained therefore that a line in any natural or artificial object points 10 or 20 or any number of degrees north or south, or above or below, &c. the centre of a scene or picture, that is to say, the point of sight or principal visual ray, then also is it known that all the parallels to that line have their vanishing point in that spot of the field of view, and a line supposed drawn from the eye to the heavens, or really drawn to the picture in that direction, marks the true vanishing point.

It is explained now, why in a long arched tunnel,

or a cathedral with many longitudinal lines on its floor, walls, roof, &c., all such lines seen by an eye looking along from one end, appear to converge to a point at the other, like the radii of a spider's web; and why in the representation of a common room, viewed from one end, all the lines of the

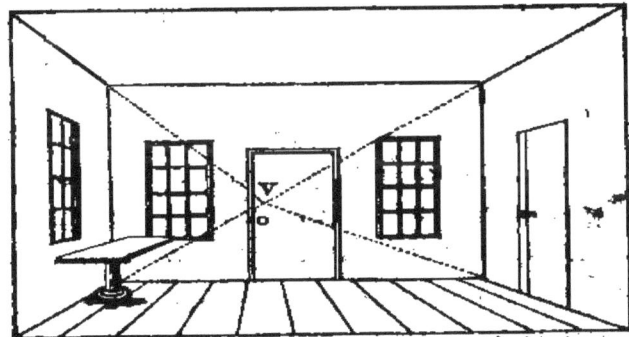

corners, tops and bottoms of windows, floor, stripes on a carpet, corners of tables, &c. being parallel to each other, tend to the same vanishing point as V, and are cut off according to the rule of foreshortening formerly alluded to. The same considerations will explain the appearance often to be observed, of two little clouds near each other and almost motionless for a long time in the distant sky directly to windward, but which on approaching the spectator, appear to be gradually, and at last suddenly enlarged, while one of them sweeps past considerably to the right hand, and the other considerably to the left, but both again meet, when at the former distance beyond the spectator, appearing there as small as at first. Clouds being so mutable and uncertain in their forms, persons have been led to deem all apparent changes in them, of form, size, and place, to be real changes, and not, as they generally are, mere optical or perspective illusion.

By far the most important vanishing point in

common scenes is the middle of the horizon or level line, and in a picture properly placed it is at the exact height of the eye. It is marked S in the figure before the last, and V in the last figure. Because in houses, the roofs, foundations, floors, windows, &c., are all horizontal, the vanishing points of their lines must be somewhere in the horizon, and if the spectator be in the middle of a street or of a building, and be looking in the direction of its walls, their vanishing point will be in the centre of the scene or picture, if he be elsewhere, it will be at one side. In holding up a picture-frame, though which to view a scene suitable for a picture, it would be found most befitting to raise it until the line of the horizon appeared to cross at about one-third from the bottom :—this fact becomes the reason of the rule in painting, so to place the horizontal line. In beginning a picture, this line is usually the first line drawn on the canvas, as marking the place of the vanishing points of all level lines and surfaces. And the eye of the spectator is supposed to be placed before the middle of it, and generally about as far from the picture as the picture is itself long, such being the extent of view which the eye at one time most conveniently commands.

Understanding now that the apparent or perspective direction of all lines in a scene is towards their vanishing points as above discovered, or parallel to the picture where the originals are so and therefore cannot have vanishing points according to the rule given, we proceed to shew how much of a line drawn to any vanishing point belongs to the known magnitude of any object through which it passes ; in other words, how much an object is in perspective foreshortened in consequence of its obliquity of position.

If we suppose A S P to represent a plate of glass seen edgeways, and that towards the point S in it, an eye is looking from the point D, evi-

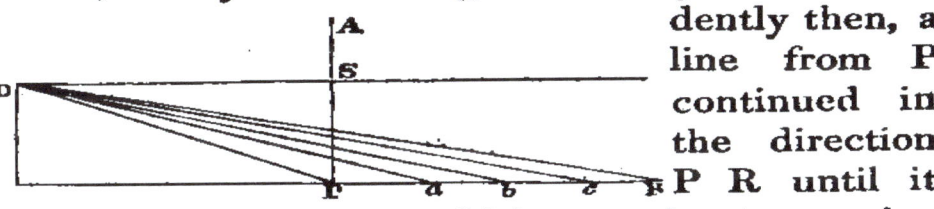

dently then, a line from P continued in the direction P R until it vanished from sight, could have as its perspective image or representation on the glass only a line reaching from P to S, the *point of sight* here, and the pictorial vanishing point of the line. Now to divide the *representative* line P S so as to correspond with any given portions of the *original* line P R, &c., it would only be necessary to draw other lines from the place of the eye D to cut or touch the original line in the situations desired, and these lines would cut the perspective line S P as required : for instance, the portion of true line *a b* would be represented by the portion of the image-line S P, included between the two lines *a* D, and *b* D, and so of any other portions. There are figures drawn on many mathematical scales by which such problems as this can be solved at once ; and the proportions are also detailed in common *tables :* but the most generally convenient mode in practice is to set off on the intended drawing (as that of which *c d* here marks the boundary), from the point of sight S a distance on the horizontal line, at D, equal to the distance of the eye from the pic-

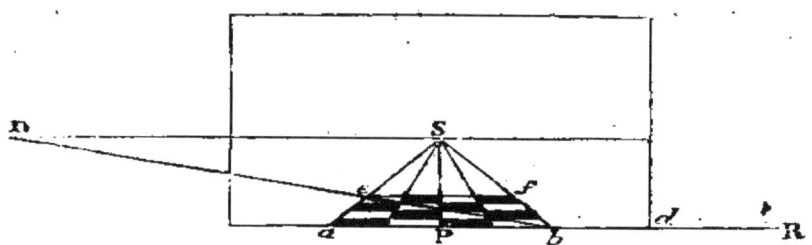

ture, and then by oblique lines drawn upon the base-
line P R, to cut the perpendicular line P S in the
situations desired—as is seen in the last sketch,
which differs from the present only in having the
point of distance marked *before* its point of sight,
instead of *laterally* as here. And the line P S
being always cut by the oblique line from D in
proportion to the length of base-line between P
and the extremity of the oblique line, a horizontal
line drawn through any point in it cuts in corresponding proportions all the other lines which have
their vanishing points in the horizontal line S D,
for instance, *a* S, *b* S, &c. Thus, to draw in perspective, on the surface above represented and
prepared, a chess board or board of squares, it is
necessary to set off the breadth of the board on the
base-line to the right and left of P, *viz.* to *b*
and *a*, and then to draw to the point of sight as a
vanishing point, the lines *a* S and *b* S, part of
which lines will therefore represent the sides of
the board, and then to draw the diagonal *b* D,
which for the reasons above stated will cut the
lines P S and *a* S in proportion to the length of
base-line to the right of their extremities ; *a e f b*
therefore is a square seen in perspective, and any

number of smaller included squares are made by drawing lines from the vanishing point to equal divisions on the base, and making cross lines where the diagonal cuts these.

Much of the delight which the art of painting is calculated to afford is lost to the world, because persons in general know not how to look at a picture. Unless a spectator place himself where he can see the objects in true perspective, so that he may fancy himself looking at them through a window or opening, every thing must appear to him falsely and distorted. The eye should be opposite the *point of sight* of the picture, and therefore on a level with the line of the *horizon,* and it should be at the required distance, which is generally at least as great as the length of the picture. But blame not unfrequently rests also with the artist, from his having neglected the study of perspective. It is very common, for instance, to see miniature resemblances of architectural structures so foreshortened and tapered, that the eye, to see them in true perspective, would require to be within an inch of the paper ; whence at the usual distance of ten or twelve inches they are seen as hideous distortions. The specimens in the few preceding pages necessarily exemplify in a degree this error, because the *point of distance* had to be marked where there was but a small page. These figures therefore, by any person studying the subject particularly, should be drawn on such a scale as that the eye may really view them at the distance supposed.

A means of judging of the dimensions of bodies by the visual angle, but which depends neither on the absolute size of the image, nor on the foreshortening of the ground plane on which the body stands, is to use known objects in view as measures for others near them which are unknown.

If any person of our acquaintance be standing at some distance from us near another person who is a stranger, we know how tall the stranger is by taking the acquaintance as a measure.

In pictorial representations of objects little familiar, as to many people are the Egyptian pyramids, the bodies of the whale, the elephant, the camel, &c., human beings may be represented around them to serve as measures for the less known object. The Colossus of Rhodes seen from afar, might to a stranger have appeared but an ordinary statue of a man, but the exact magnitude would have been known as soon as a ship of known dimensions were seen sailing into port between his gigantic limbs.

When an unpractised eye is first directed to a great ship of war, it will on many accounts dwell upon it with wonder and admiration; but it may not judge truly of the enormous magnitude until near enough to perceive the sailors climbing on the rigging, and appearing there, by comparison, as flies or little birds appear among the branches of a majestic tree.

By having a measure of this kind presented to us, the magnitude and elevation of some fine edifices are rendered more obvious. The magnificent pile of St. Paul's in London becomes more striking still, when we discover visitors looking from the balconies near the summit cross. They appear so minute among the surrounding huge masses that a person is at first for a while disposed to doubt whether they be men; but the fact once ascertained, the grandeur of the temple is rendered extremely impressive.

Many persons cannot distinguish between the little pilot balloon (sometimes despatched before a great one to shew the direction of the wind) and the great balloon itself, until with the last they perceive the aeronauts as little black points suspended under the globular cloud.

Strangers visiting Switzerland, on first entering the vallies there, are often much deceived as to their extent. Because familiar generally with more lowly hills and shorter vallies at home, but which from being near to the eyes form bulky images, and having no measure at first, they almost universally underrate the alpine dimensions:—they will wonder, for instance, in the valley of Chamouny, that they should be travelling swiftly for hours without reaching the end, where on entering they did not believe the length to be as many miles.

The author once sailed through the Canary Islands, and passed in view of the far-famed Peak of Teneriffe. It had been in sight in the afternoon of the preceding day, at a distance of more than 100 miles, appearing then only as an ordinary distant hill rising out of the ocean; but next morning, when the ship had arrived within about twenty miles of it, and while another ship of the fleet, holding her course six miles nearer to the land, served as a measure, it stood displayed as perhaps the most stupendous single object which on earth, and at one view, human vision can command. That noble ship, whose side shewing its tiers of cannon, equalled in extent the fronts of ten large houses in a street, and whose masts shot up like lofty steeples, appeared as a speck scarcely rising from the sea, compared with the huge prominence beyond it towering sublimely to heaven, and around which the masses of cloud, although as lofty as those which sail over the fields of Britain, were still hanging low on its sides. Teneriffe alone of very high mountains, rises out of the bosom of the ocean with inaccessible steepness on one side, to an elevation of 13,000 feet; and as an object of contemplation, therefore, is more impressive than even the still loftier summits of Chimborazo or the Himalayas, which rise from elevated plains, and in the midst of surrounding hills.

It is because objects which are nearly on a level with us, as contrasted with such as are either much above or much below, are in general more numerously surrounded by other objects which serve as measures of comparison, that we judge so much more correctly of the size and distance of the former than of the others.

A man walking like ourselves on the sea-shore or other level, is at once recognized; and probably it may not occur to us, that he appears smaller on account of the distance; but if the same man be seen afterwards at an equal distance above us, collecting the sea-fowl's eggs on the face of a cliff, or below us, gathering shells on the beach when we ourselves have reached the height, he appears no bigger than a crow: yet in all the cases he is where the same bulk forms the same magnitude of image on the retina.

Even on a horizontal plain, if the general surface be bare and uniform, single distant objects appear very diminutive. This is true, for instance, of a man seen apart from his caravan, while journeying across a sandy desert; while a man viewed at an equal distance, in the midst of a cultivated landscape, appears of his natural size; the same is true of a boat or ship seen out on the high sea, as contrasted with similar objects viewed in a harbour, where other known objects are near them.

We may now understand why the sun and moon, when rising or setting, appear to us much larger than when they have attained meridian height—although, if we examine them by any measure of the visual angle, as simply by looking at them through the same ring or tube, we find that there is no difference. The sun and moon in appearance from this earth are nearly of the same size, *viz.* always occupying in the field of view about the half of a degree, or as much as is occupied by a circle of a foot in diameter when held about 250 feet from the eye—which circle therefore at that distance, and at any time, would just hide either of them. Now when a man sees the rising moon apparently filling up the end of a street, which he

knows to be 100 feet wide, he very naturally believes that she then subtends a greater angle than
usual, until the reflexion occur to him,—which it
rarely will of itself, that he is using as a measure
of her size, a street known indeed to be 100 feet
wide, but of which the part concerned, owing to
its distance, appears to his eye exceedingly small.
The width of the street near him may occupy 60°
of his field of view, and he might see from between the houses many broad constellations instead of the moon only, but the width of the
street far off may not occupy, in the same field
of view, the twentieth part of a degree, and the
moon, which always occupies half a degree, will
there appear comparatively large. The kind of
illusion now spoken of is yet more remarkable
when the moon is seen rising near still larger
known objects,—for instance beyond a town, or a
hill which then appears within her luminous circle.
Any person who from the river-side terraces of
Greenwich has observed the sun setting beyond
London, with St. Paul's cathedral included in the
glorious picture, will recollect a most interesting
example of our present subject. That our ocular
judgment of the size of the sun or moon is thus
influenced by the presence or absence of objects
of comparison, and not by the place of the bodies in the sky, is proved by the fact that a person
viewing these bodies from the bottom of some of
the Swiss valleys, where he might almost suppose
himself placed at the centre of the earth, and
looking abroad along an endless extent of precipices—if he can closely compare them with cer-

tain known magnitudes of ridge or forest bounding his view, sees them, although at a great elevation, as large as they appear from other situations when rising directly out of the sea. Another proof is afforded by the case of a balloon at a great elevation seen crossing the disk of the sun or moon, and appearing, however large, as an absolute speck within the vast luminous area. In a future paragraph it will be explained, that the sun and moon when low appear larger than when high, also, because of their apparent dimness when low.

It may be remarked here, that the visual estimate formed of the great size of the sun and moon when seen on the horizon, is not an illusion, as is popularly supposed, but an approximation to truth, still vastly short of the reality. When we see a tree or a house, or a hill, apparently within the circumference of one of these orbs, it is really true that the orb is larger than the tree, or house, or hill, and so much larger, that even if the whole British Isles could be lifted away from the earth, and suspended as a map in the sky, when brought near the moon they would hide from a spectator on earth but a small part of her.

Having now shewn that in relation to any object, the visual angle or apparent size can be a measure of the true size only when the distance also is known, and that the visual angle itself serves to determine the distance only in certain cases, we proceed to examine other means which the eye commands for guessing at distances.

2d. Intensity of light, shade, and colour. (See the Analysis, page 162.)

It has already been explained that light, like every other influence radiating from a centre, becomes rapidly weaker as the distance from the centre increases, being, for instance, only one fourth part as intense at double distance, and in a corresponding proportion for other distances; while it is still farther weakened by the obstacle of any transparent medium through which it passes. Now the eye soon becomes sufficiently familiar with these truths to judge from them, with considerable accuracy, of the comparative distances of objects.

The fine gothic pile of Westminster Abbey may break upon the view in some situation where nearer edifices, and perhaps some minor imitations of its beauties, already fill and dazzle the eye with their brightness, but the misty or less distinct outline of the former warn the approaching stranger of its true magnitude, and prepare him for the enjoyment which a nearer inspection of its grandeur and perfection is to afford.

A small yacht or pleasure-boat may be built from the same model or of the same comparative dimensions as a first rate vessel of war, and may be in view from the shore at the same time, only so much nearer than the ship, that both shall form images of the same magnitude on the retina of a spectator. In such a case, to an unpractised eye, it might be difficult to detect the difference, but to another, the bright lights of the little vessel, contrasted with the softer or more misty appearance of the larger, would leave no room for doubt. A haziness occurring in the atmosphere between the little vessel and the eye, might considerably disturb the judgment.

In a fleet of ships, if the sun's direct rays fall upon one here and there through openings among the clouds, while the others remain in shade, the former immediately start in appearance towards the spectator. Similarly, the mountains of an unknown coast, if the sun-shine fall upon them, appear comparatively near, but if clouds again intervene, they recede and mock the awakened hope of the approaching mariner.

A conflagration at night, however distant, appears to spectators generally, as if very near, and inexperienced persons often run towards it with hope of arriving immediately, but find after miles travelled that they have made but a little part of their way.

A person ignorant of astronomy deems the heavenly bodies vastly nearer to the earth than they are, merely because of their being so bright or luminous. The evening star, for instance, seen in a clear sky over some distant hill-top, appears as if a dweller on the hill might almost reach it—for the most intense artificial light that could be placed on the height would be dim in comparison with the beauteous star, yet to a dweller on the hill it appears just as distant as to one on the plain ; nay, at thousands of miles nearer, the appearance would still be nearly the same.

The concave of the starry heavens appears flattened above, or nearer to the earth in its zenith than towards its horizon, because the light from above having to pass through only the depth or thickness of the atmosphere, is little obstructed, while of that which darts towards any place horizontally through hundreds of miles of dense vapour-loaded air, only a small part arrives.

The sun and moon appear larger at rising and setting than when midway in heaven, partly, as already explained, because while below they can easily be compared with other objects, of which the size is known, but partly, also, because of their less light in the former situation, while their diameters are always the same.

A fog or mist is said to magnify objects seen through it. The truth is, that by reason of the diminished intensity of light, it makes them appear further distant without lessening the visual angles subtended by them; and because an object at two miles, subtending the same angle as an object at one mile, must be twice as large, the conclusion is drawn that the dim object is large. Thus a person in a fog may believe that he is approaching a great tree, fifty yards distant, when the next step throws him into the bush which had deceived him.——Two friends meeting in a fog have often mutually mistaken each other for persons of much greater stature.——A row of foxglove flowers on a neighbouring bank has been mistaken for a company of scarlet-clad soldiers on the more distant face of the hill. There are, for similar reasons, frequent misjudgings in late twilight and early dawn.——The purpose and effect of a thin gauze screen interposed between the spectators in a theatre and some person or object meant to appear distant, are intelligible on the same principle; a boy near, so screened, will appear to be a man at a distance.—The art of the painter uses sombre colours when his object is to produce in his picture the effect of distance.—On the alarming occasion of a very dense fog coming on at sea, where the ships of a fleet are near to each other, without wind, and where there is considerable swell or rolling of the sea, much damage is often done, but it is to be remarked in such a case that the size of ships approaching to the shock is always in idea exaggerated.

The celebrated *Spectre of the Brocken*, among the Hartz Mountains, is a good illustration of our present subject. On a certain ridge, just at sunrise, a gigantic figure of a man had often been observed walking, and extraordinary stories were related of it. About the year 1800 a French philosopher went with a friend to watch the phenomenon; but for many mornings they had paraded on an opposite ridge in vain. At last, however, they discovered the monster, but he was not alone; he had a companion, and singularly he and his companion aped all the motions and attitudes of the observer and his companion: in fact, the spectres were merely shadows of the observers, formed by the horizontal rays of the rising sun falling on the morning fog which hovered over the valley beyond; but because the shadows were very faint, they were deemed distant, and therefore seemed men walking on the opposite ridge, and because a comparatively small figure seen near, but supposed distant, appears of gigantic dimensions, these shadows were accounted giants.

While under common circumstances the comparative intensities of light furnish an indication of difference of distance, there are other cases of comparative intensity, in which bodies are to be considered not as wholes, but as made up of parts unequally exposed to the source of light, and therefore reflecting it to the eye, or being illuminated in different degrees. In observing, for instance, a white house exposed to the sun, it is seen that the side directly receiving the rays is highly illuminated or bright, while the other sides are less so, and are said to be in the shade : and they are luminous in proportion to other sources of reflected light near them. The different faces or walls of such a house are as strongly distinguished from each other, by the mere difference of shade, as if they were of different colours. If the object were a ball instead of a square house, there would still be as great differences of shade in the half not receiving direct rays, but the parts, instead of forming abrupt contrasts like the walls of a house, would melt into each other and mark the beautiful round contour of the object. The consideration of all such cases forms the subject of chiaro-oscuro, so interesting to the painter.

Had there not been in nature the provision of light and shade now described, the sense of sight would have been of comparatively little use, for it is that provision chiefly which enables us to distinguish the profile or outlines of different bodies placed near to each other, and the protuberant or other form of the surfaces next the observer. But for this, it would have been impossible to distinguish, for instance, between a flat circle, a sphere, and a cone, all directly exposed to the eye; but, in reality, the uniformly bright surface of the circle, the soft rounded shadowing of the sphere, and the shade coming to a point on the cone, at once declare the true forms. But for the shadows, the façade of a white palace of varied architecture would have been an unmeaning sheet of light; the lights, however, and shadows produced by the juttings and recesses, mark the variety of surface most completely; and the round pillar is distinguished from the square, and every pediment, and capital, and architectural ornament, stands out pleasingly conspicuous. But for light and shade, again, the human face divine would have been an unmeaning patch of flesh, for there are few other lines in it than those made by different exposures to the light, and yet every prominence and depression are so truly indicated to the eye that it becomes full of meaning or expression. How clearly mere light and shade serve to convey all that the eye can learn of a scene or object, may be perceived by examining any of the admirable engravings which now abound—scarcely inferior in expression to the most finished paintings.

The student of painting soon learns that the lines called outlines, by which he first sketches subjects, do not exist at all in nature, and have to be again effaced in his finished work: for they only mark the place where lights and shades happen to meet. Much may be conveyed to the mind however by a mere outline, and particularly if lines of different breadth or thickness are used to mark the situation of the lighter and deeper shadows.

The subject of chiaro-oscuro is not so simple as, from the fact of the sun being the great source of light, might at first be supposed; for although this be true, still every body which reflects the sun's light becomes a new source to those about it, and the shading of a picture must have reference to all such sources, and to the colours of the body itself, and of the neighbouring bodies.

In looking at an extended landscape, it is seen that the near objects considered as wholes, are comparatively bright, that their shadows are strongly marked, and that their peculiar colours are every where easily distinguishable — as of flowers, fruit, foliage, &c., but farther off the colours become dim, the lights and shadows melt into each other or are confused, and the illumination altogether becomes so faint that the eye at last sees only an extent of distant blue mountain or plain—appearing blueish because the transparent air through which the light must pass has a blue tinge, and because the quantity of light arriving through the great extent of air is insufficient to exhibit the detail. The ridge called Blue Mountains in Australia, and another of the same name in America, and many others elsewhere, are not really blue, for they possess all the diversity of scenery which the finest climates can give, but

to the discoverer's eye, bent on them from a dis
tance, they all at first appeared blue, and they
have ever since retained the name.

In a picture by an artist, who on his canvas
stretched on a frame disposes the lights, shades,
and colours in the very situations and with the
intensities which on coming from the landscape to
the eyes, through a plate of glass filling up the
frame, they would have had, all that we have now
been saying is strictly exemplified. In the foreground the objects are large and bright, but as the
scene is supposed to be gradually more remote,
the size and brightness of the objects correspondingly diminish, until at last there is only a dim
mixture of blueish or greyish masses forming the
horizon and sky.

A child, during what may be called the education of the sense of sight, has a strong perception
of the vast differences of appearance which things
assume according to their accidental distance from
the eye, their position, their exposure to light,
&c. ; for these differences, being often calculated
to deceive the young judgment, many of them
have been noted with surprise. Thus a boy when
he first discovers that a ship which at the quay,
with sails spread, concealed from him half the
heavens, is in an hour or two afterwards seen by
him on the distant horizon as a speck hardly big
enough to hide one star ;—or again, when he discovers that the faint blue unchanging mass which
he had always observed bounding in one direction
the view from the home of his infancy, is a distant mountain-side thickly inhabited, and covered

with fields and gardens, where in succession all
the bright colours of the different seasons predo-
minate—his attention is strongly awakened, and
he feels surprise.   But as soon as experience has
enabled him to interpret readily and correctly the
visual signs under every variety of circumstance,
his attention passes so instantly from them to the
realities which alone are interesting to him—just
as it might pass from the paper and printing of a
newspaper to the important intelligence commu-
nicated by them, that he very soon ceases to be
aware that the sign, which in every case similarly
suggests the object, is not also in every case similar
to itself, and the very same true and complete representation of the reality.   The prejudice that
the sign is of this nature, becomes quickly so
strong, that even a difficult effort has to be made
by a grown person again to attend to the mere
appearances, in any scene of which the *realities* are
known.

This attempt to analyze the appearances,
and which in one sense is a trying to unlearn
something, or to retrograde, is called the study of
perspective—and when it regards the apparent re-
duction of size, and the foreshortening of bodies
under various circumstances, it is called *linear
perspective ;* when it regards the fading of light
and the modifying of colour, it is called *aerial
perspective.* As the art of painting depends entirely upon the understanding of these two departments, the gradual progress which it has made
in different countries is a measure of the degree
in which the common prejudice that things *appear*

as they *are* has in them been overcome. Where this prejudice exists, any untaught person conceives a good painting to be merely a miniature representation drawn according to a certain reduced scale,— as of an inch to a yard,—and in which all the dimensions of things are to be measured as simply as in the reality—while the colours as to vividness, &c. should perfectly agree in both. This statement is remarkably illustrated by the fact, that children in their rude attempts to paint always aim at realizing the notion of the art given above, and that such has been the first stage of painting in every country. In Europe now, owing to the labours of men of genius, art in painting may be said almost to rival nature, producing scenes as lovely as the finest of nature's scenes, and scarcely distinguishable from them, but in other countries, as in China and India, among the native artists, the first stage of the art is still in existence. In a Chinese picture, owing to the absence of perspective proportions, an extensive subject is only a collection of portraits of men and things drawn on the same scale, and placed near one-another, and where all the colours are as vividly shewn as if the objects were only a few feet from the eye : there the figures at the bottom or foreground are supposed to represent the objects nearest to the spectator, while the figures higher up are supposed to be of more remote objects, all appearing as they might be seen in succession by a person who had the power of flying over the country. This kind of picture or representation, although not natural if all viewed at once, may communicate more information than a single common painting, for it is equivalent to many such. In Europe, and lately, the principle has been usefully acted upon for certain purposes, as for representing on one long sheet or on a succession of sheets connected in a suitable manner, the banks of a river or a line of road. The banks of the Rhine particularly have thus been admirably pourtrayed, so that the spectator directing his eye along the paper, feels almost as if carried in a balloon to view in great detail the whole of the real and enchanting scenery. The principle might perhaps with advantage be acted upon still more extensively—for instance, to produce, instead of common maps or charts of countries, true bird's-eye views, over which the eye moving from place to place, at every new point of sight, would see a certain portion of the country just as a bird or aeronaut would, the sketch being supposed to be taken from that certain elevation deemed most suitable for the ends in view.

3d. Divergence of the rays of light.

This is the next circumstance to be mentioned by which the eye judges of distance. Supposing the line E F to mark the place or breadth of the pupil of the eye, the light entering from an object at *a* which is near, is very divergent, or is

spreading with a large angle ; from *b* the extreme rays are less divergent, or they open at a smaller

angle ; from *c* they are less divergent still, and so
on. Now the eye to form an image on its retina
requires to exert a bending power exactly proportioned to the divergence of the rays ; and it
appears to have a sense of the effort made, which
becomes to the person a kind of measure of the
distance of the object. This great divergence of
the rays entering the eye, is the chief circumstance
in which the most perfect painting must still differ
in its effect upon the eye from a natural scene—
for while in nature every object according to its
distance is sending rays which reach the eye with
different divergence, and which therefore can produce distinct images on the retina at any one time
only of the objects which are at due distance from
it, the rays from a picture which is a single plane
surface, come from every part with the same divergence, and the eye must feel a disappointment
in not having to accommodate its power of bending to the different distances attempted to be pourtrayed on the canvas. It might be expected that
this kind of disappointment would be more felt on
looking at a common picture placed a few feet
from the eye, than at the sort of picture called
panorama, which is on a larger scale and proportionately more distant, but such is not the case ;
and the reason seems to be, that in the former the
illusion is not intended to be complete, the fact
of its being but a picture not being at all concealed, and the eye is therefore at once told to
expect a difference of feeling;—but in the panorama, the whole circumstances are arranged to deceive the eye entirely, if possible, and to make it

believe that the images on the retina are formed by light from the objects themselves; then to the eye really deceived in all other particulars, the non-accordance with nature in this one is strongly, and by some persons even painfully felt, so as on their first entering the place to cause them headach or giddiness.—The illusion, and consequently the pleasure from viewing any picture, may be made more complete by the spectator using lenses or spectacles, such that the focal distance shall be equal to the distance of the painting from the eye; because such lenses, as was formerly explained, would render all the rays entering the eye nearly parallel, and therefore very nearly such as would arrive from objects at any considerable distance.

4th. Convergence of the axes of the eyes.

This is the last circumstance to be mentioned, by which a person, through the eye, judges of the distance of objects. In consequence of there being in the two eyes corresponding parts which must be similarly affected by any object, that the person may have single vision of it—as was explained in a former page, the axes of both eyes must point to the object, and if it happen to be very near, they will meet and cross each other so near the face as to produce the appearance of squinting,—seen when a person tries to look at the point of his nose—but if the object be more distant, the obliquity will be less, until at last the eyes directed to a thing at a very great distance, will have their axes almost parallel. The last figure may serve also to explain this subject. Supposing E and F to mark the places of the two eyes, if the object looked at be near them, as at *a*, they must be very much turned inwards, that their axes may meet; if it be at *b*, they will be less turned, if at *c* less still, and so forth.

When the eyes are not directed to any thing in particular, the axes generally become parallel, or as if they were pointed to a very distant object: and because this happens generally when persons are reflecting on things which are absent and seen only by the mind's eye, it is an expression of countenance held to mark contemplation or thoughtfulness.

The direction of the visual axes is a particular like the divergence of light, as to which a mere picture can never produce upon the eye precisely the effect of the objects themselves. To see a picture the axes must meet at a few feet from the eye; while to see the objects of nature, they often do not meet nearer than at miles. By a glass, however, as will be explained a little further on, it is possible to correct also this defect, and to render an optical illusion, as regards still objects, almost complete.

When a picture has to represent objects supposed far from the eye, the farther the picture itself is placed from the eye, supposing the figures to be made proportionately large, the more nearly perfect will the illusion become, because the divergence of rays and convergence of the axes (the two circumstances in which the effect on the eye of a mere picture must always differ from that of

T

a real scene) will be in proportion more nearly natural. This explains in part why the picture called panorama (from Greek words, signifying a *view* of *all*) is an exhibition so charming; for usually the painting is far removed from the eye, and is drawn on a proportionately large scale, and the eyes feel that the light comes from a considerable distance, and that their axes do not need to converge very much; and when, in such a case, the first impression of the want of perfect conformity has passed away, the illusion becomes nearly complete. But a not less important peculiarity in the panorama is, that instead of being a painting on a plane surface like common pictures, and embracing only a small part of the field of view, it is on a surface which entirely surrounds the spectator, and on which all the objects visible in every direction from the supposed place, are seen in the very situations which in nature they hold; and the spectator is enabled to conceive much more distinctly of each particular by seeing it in relation to others around. Few persons can forget the impression made on them by the first panorama which they may have seen; and with increased maturity of judgment, still more and stronger reasons are discovered for admiring this miraculous mode of instantly transporting persons to any distance, beyond seas and other dangers, to contemplate at their ease the most interesting scenes of nature, represented under the most favourable circumstances of light, season, weather, &c. Hence few persons of good taste neglect the opportunity now, in most great towns so frequently offered, of obtaining at so little cost so high a gratification.

To correct the slight remaining optical defects of a common panorama, a large lens may be used, of which the focal distance is the distance of the picture from the eye. This has the effect of converting the divergence of the rays of light into the parallelism which belongs to the supposed remoteness of the objects, and it also bends the light so that the axes of the eyes become parallel. The author has found a convenient mode of using the lens for such a purpose to be to cut out two round pieces from opposite sides of it, and to form them into a pair of spectacles:—from one lens three pairs may be formed. Panorama exhibiters should always keep such lenses or spectacles for the use of visitors.

The effect of the magnitude and distance of the ordinary large panoramic views might be had with the assistance of proper glasses, from even the smallest picture or engraving embracing the same field; and it is remarkable that some enterprizing person has not undertaken to publish a series of interesting views fitted to be used in that way. A common panorama occupying a circular wall of 150 feet circumference and twenty feet high, might be reduced, still retaining the same truth of proportions, to appear on a sheet of paper five feet long and eight inches high or broad—or on a common sheet of paper, which might afterwards be cut into three stripes to be joined endways; and if this paper were set up in a suitable frame, like a wall round the head of a spectator, while its edges were concealed by drapery or otherwise, and the eye could only view it through fit glasses placed in its centre and made to turn round so as to command the whole, it could not by an ordinary spectator be distinguished from the large panorama. With the art of lithography, now so well adapted to producing soft representations of scenery, the expense of such views might be very moderate, allowing them to form a common part of library furniture. When we reflect upon the expansion of mind obtained by travelling, and that not a few of the advantages would follow a familiarity with a good selection of panoramic views, it is not perhaps too much to suppose that courses of instruction on geography, history, &c. may before long be illustrated by this most interesting mode of aiding the conception and memory.

Common paintings and prints may be considered as parts of a panoramic representation, shewing as much of that general field of view which always surrounds a spectator, as can be seen by the eye turned in one direction, and looking through a window or other opening. The pleasure from contemplating these is much increased by using a lens or such spectacles as above described. There is such a lens fitted up in the shops, with the title of *optical pillar machine*, or *diagonal mirror*, and the print to be viewed is laid upon a table beyond the stand of the lens, and its reflexion in a mirror sup-

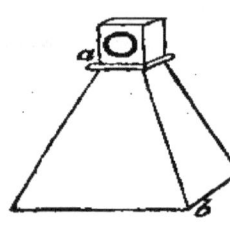

ported diagonally over it, is viewed through the lens. The illusion is rendered more complete in such a case by having a box, as *a b*, to receive the painting on its bottom, and where the lens and mirror, fixed in a smaller box above at *a,* are made to slide up and down in their place to allow of readily adjusting the focal distance. This box used in a reverse way becomes a perfect camera obscura. The common shew-stalls seen in the streets are boxes made somewhat on this principle, but without the mirror ; and although the drawings or prints in them are generally very coarse, they are not uninteresting. To children whose eyes are not yet very critical, some of these shew-boxes afford an exceeding great treat.

A still more perfect contrivance of the same kind has been exhibited for some time in London and Paris under the title of *Cosmorama* (from Greek words signifying *views* of the *world,* because of the great variety of views). Pictures of moderate size are placed beyond what have the appearance of common windows, but of which the panes are really large convex lenses fitted to correct the errors of appearance which the nearness of the pictures would else produce. Then by farther using various subordinate contrivances, calculated to aid and heighten the effects, even shrewd judges have been led to suppose the small pictures behind the glasses to be very large pictures, while all others have let their eyes dwell upon them with admiration, as magical realizations of the natural scenes and objects. Because this contrivance is cheap and simple, many persons affect to despise it ; but they do not thereby shew their wisdom : for to have made so perfect a representation of objects, is one of the most sublime triumphs of art, whether we regard the pictures drawn in such true perspective and colouring, or the lenses which assist the eye in examining them.

It has already been stated, that the effect of such glasses in looking at near pictures, is obtainable in a considerable degree without a glass, by making the pictures very large and placing them at a corresponding distance. The rule of proportion in such a case is, that a picture of one foot square at one foot distance from the eye, appears as large as a picture of 60 feet square at 60 feet distance. The exhibition called the *Diorama* is merely a large painting prepared in accordance with the principle now explained. In principle it has no advantage over the cosmorama or the shew-box, to compensate for the great expense incurred, but that many persons may stand before it at a time, all very near the true point of sight, and deriving the pleasure of sympathy in their admiration of it, while no slight motion of a spectator can make the eye lose its point of view.

A round building of prodigious magnitude has lately been erected in the Regent's Park in London, on the walls of which is painted a representation of London and the country around, as seen from the cross on the top of St. Paul's Cathedral. The scene taken altogether is unquestionably one of the most extraordinary which the whole world affords, and this representation combines the advantages of the circular view of the panorama, the size and distance of the great diorama, and of the details being so minutely painted, that distant objects may be examined by a telescope or opera-glass.

From what has now been said, it may be understood, that for the purpose of representing still-nature, or mere momentary states of objects in motion, a picture truly drawn, truly coloured, and which is either very large to correct the divergence of light and convergence of visual axes, or if small, is viewed through a glass, would affect the retina exactly as the realities. But the desideratum still remained of being able to paint motion. Now this too has been recently accomplished, and in many cases with singular felicity, by making the picture transparent, and throwing lights and shadows upon it from behind. In the exhibitions of the diorama and cosmorama there have been represented with admirable truth and beauty such phenomena as—the sun-beams occasionally interrupted by passing clouds, and occasionally darting through the windows of a cathedral and illuminating the objects in its venerable interior—the rising and disappearing of mist over a beautiful landscape,—running water, as for instance, the cascades among the sublime precipices of Mount St. Gothard in Switzerland;—and, most surprising of all, a fire or conflagration. In the cosmorama of Regent Street, the great fire of Edinburgh was admirably represented:—first, that fine city was seen sleeping in darkness while the fire began, then the conflagration grew and lighted up the sky, and soon at short intervals, as the wind increased, or as roofs fell in, there were bursts of flame towering to heaven, and vividly reflected from every wall or spire which caught the direct light—then the clouds of smoke were

seen rising in rapid succession and sailing north-
ward upon the wind, until they disappeared in the
womb of distant darkness. No one can have
viewed that appalling scene with indifference, and
the impression left by the representation, on those
who knew the city, can scarcely have been weaker
than that left on those who saw the reality. The
mechanism for producing such effects is very
simple ; but spectators, that they may fully enjoy
them, need not particularly inquire about it.

It is remarkable, when the imagination is once
excited by some beautiful or striking view, how
readily any visual hint produces clear and strong
impressions. One day in the cosmorama, a school-
boy visitor exclaimed with fearful delight that
he saw a monstrous tiger coming from its den
among the rocks ;—it was a kitten belonging to
the attendant, which by accident had strayed
among the paintings. And another young spectator was heard calling that he saw a horse gallop-
ing up the mountain side ;—it was a minute fly
crawling slowly along the canvas. There is in
this department a very fine field yet open to the
exercise of ingenuity, for the contemplation of
pictures representing motion or progressive events,
may be made the occasion of mental excitement
the most varied and intense. For instance, there
are few scenes on earth calculated to awaken more
interesting reflections on the condition of human
nature than that beheld by a person who sails
along the river Thames from London to the sea,
a distance of about sixty miles, through the wonders which on every side there crowd on the

sight—the forest of ships from all parts of the world—the glorious monuments of industry, of philanthropy, of science—the marks of the riches, the high civilization, and the happiness of the people. Now this scene was last year in one of our theatres strikingly pourtrayed by what was called a *moving panorama* of the southern bank of the Thames. It was a very long painting, of which a part only was seen at a time gliding slowly across the stage, and the impression made on the spectators was, that of their viewing the realities while sailing down the river in a steam boat. In the same manner the whole coast of Britain might be most interestingly represented—or any other coast, or any line of road, or even a line of balloon flight. There was another *moving panorama* exhibited about the same time at Spring Gardens, aiming at an object of still greater difficulty, *viz.* to depict a course of human life; and the history chosen was that of the latter part of Buonaparte's career. Scenes representing the principal events were, in succession, and apparently on the same canvas, made to glide across the field of view, so designed that the real motion of the picture gave to the spectator the feeling of the events being only then in progress, and with the accompaniments of clear narration and suitable music, they produced on those who viewed them the most complete illusion. The story began with the blow struck at Buonaparte's ambition in the battle of Trafalgar, and to mark how completely, by representations of various moments and situations of the battle, the spectators were in imagination made present to it, the author of this work may mention, that on the occasion of his visiting the exhibition, a young man seeing a party of British preparing to board an enemy's ship, started from his seat with a *hurra*, and seemed quite surprised when he found that he was not really in the battle. To the first views there succeeded many others, similarly introduced and explained, in each of which the hero himself appeared : there were, his defeat at Waterloo—his subsequent flight—his delivery of himself to the British Admiral—his appearing at the gangway of the Bellerophon to thousands of spectators waiting in boats around while he was in Plymouth Harbour, previous to his departure for ever from the shores of Europe —his house and habits during his exile, with various views of St. Helena ;—and last of all, that solemn procession, in which the bier with his lifeless corpse appeared moving slowly on its way to the grave under the willow-tree. The exhibition now spoken of might have been made much better in all respects, yet in its mediocrity it served to prove how admirably adapted such unions of painting, music, and narration, or poetry, are to affect the mind, and therefore to become the means of conveying most impressive lessons of historical fact and moral principle.

Painting, whether employed to pourtray scenes of entirely still nature, or scenes involving some kind of motion as above described, has still as its great aim or end merely to represent interesting subjects, and to give to the spectator as much as possible the clear conception of them, which is obtained by ocular examination of realities. Painting thus, as a system of visual signs of thought, becomes like language, which is a system of audible signs, a means of expanding the boundaries of individual human existence into wider space and time, and thus of ennobling human nature. While it pourtrays only strict matters of fact, whether of past or present time, as particular human individuals, objects of natural history, the beautiful and magnificent scenes of nature, interesting events which the artist had the means of faithfully representing, &c., it may be called truly historical painting, embodying the materials of true history, both natural and civil, and then is of singular value. But even when applied to other purposes, it may still be fraught with delight ; and just as language, of which the grand object or use is to express strict truths, has still been admirably employed in giving a permanent existence to a variety of fictions, from the wildest fables and rhapsodies to the historical plays and novels of modern times, as those of Shakespeare and of Scott—which plays and novels, although not furnishing true portraits of individual human nature, are yet most correct portraits of general human nature—so may painting be employed in embodying fictions adapted to its peculiar powers, and so as to prove the artist endowed with the highest degree of human genius. It should always be recollected, however, that what is usually dignified with the name of historical painting, really bears to historical truth only the kind of relation which novels and plays bear to it, and often approaches much less nearly to the truth : for it pretends to relate a thousand minute circumstances which no history has preserved, and which therefore only the imagination of the artist can supply. Thus when a painter, knowing that Lucretia stabbed herself in the presence of her father and others, after the crime of Tarquin, exhibits a woman dying, and a certain number of persons around her in horror and astonishment, he no more represents the real Lucretia and her friends than he represents any other particular young woman and her friends ; for he is quite assured that not one of the figures in such a picture is a portrait of the individual whose name it bears : his picture therefore in so far is an untruth or fiction, while it very probably shares the additional errors and even absurdities so common among historical painters—errors, for instance, in respect of national usage in costume, religion, manners, &c., and in respect to general appearance, as when a Reubens wishing to represent Sabine or other ladies, gave them the degree of corpulency deemed comely in his own country, although it notably contrasted with the true forms of Italian or Grecian nymphs. From all this it appears that historical pictures may often be regarded as portraitures, not of the realities, but of comedians acting scenes in historical plays intended to represent the realities.

In dealing with the events of ordinary history, there is no strong reason why artists may not please themselves and their spectators as we have now been describing ; but it may admit of doubt whether similar liberties should be allowed with respect to religion. Yet any painting of *the last supper*, for instance, or of the *ascension*, is not likely to be more true than a theatrical representation. To judge of the nature of such a picture we have only to suppose any of the events recorded in the New Testament to be represented by a painter in China with the countenances such as are seen on Chinese tea-boxes ; such a representation would appear in Europe revoltingly absurd ; but the common practice here is only a degree better ; as twenty painters undertaking to treat the same subject, will put different persons into all the situations. Then it can produce no pleasing impression on a Christian's mind to be told, that an admired painting of the crucifixion was made chiefly from the body of an executed murderer ; or that for a praised representation of the triumphal entry into Jerusalem, the painter had deemed his own physiognomy the most befitting for the principal figure, while he copied the portrait of a noted modern sceptic as a specimen of the bad men, of an equally noted believer as a specimen of the good, while wife, cousins, acquaintances, and old clothes-men, served to make up the remaining groups. With the knowledge that such things have often been, it need not surprise that many persons of correct feeling turn with horror from all these mimicries and falsehoods, to seek their idea of God and his providence in the sublime descriptions of his attributes, which written language conveys, and which all creation, in a mute language scarcely less impressive, so strongly confirms. When men generally could not read, and as a mass were extremely ignorant, various means of fixing their attention upon religious subjects might be useful, and therefore proper, as sacred plays, certain processions, pictures, &c., which have now in many countries ceased to be either; but a person of good sense will continue to regard with a certain respect whatever at any time may have contributed to reclaim portions of mankind from barbarism and wickedness to the just appreciation of the divine charities of a pure religion.

There are in painting other classes of fictions, which pretend to nothing beyond fiction, and which are yet truly admirable : such are personifications of the virtues and vices, serving to recommend the practice of the former, and to deter from that of the latter;—almost all Hogarth's works are of this character, and evince the highest mental acumen and genius :—then may be mentioned the personifications of what have been called the elements and powers of nature, including many of the personages of the Heathen mythology—then other generalisations of the characteristics of human or other nature, as scenes of domestic affection, of the play of the passions, &c. &c. ; and because many subjects when so sketched, are intelligible to the eye with the suddenness of lightning, where longest verbal description would convey the idea but imperfectly, the art of painting in regard to them possesses a truly magical and inestimable power.

As painting, whether engaged in representing matters of fact or of fiction, can accomplish its ends only through the arts of *drawing* or linear

perspective, and of shading and colouring, or
aerial perspective, these subjects require to be
studied by every artist with great attention ; but it
is important for all to be aware that the greatest
mastery over these, which are merely the mechanical parts of the art, will go a very short way
towards producing good performances, unless
there be present also the genius to select or to
compose subjects worthy of being represented.
The latter remark seems the more necessary, because there is in human nature a disposition to
value so much the means by which important ends
are attained, that often the end itself is forgotten in the contemplation of the means,—while
among painters, as among persons of other occupations, the talent for the inferior or more mechanical departments of the art, is more common than
for the higher. It is hence that the subordinate
accomplishments of the painter are by not a few,
both artists and pretended connoisseurs, supposed
to be the principal. But this is evidently to value
the dress or clothing, instead of the person, as
might be observed of the criticism of a mere tailor
or dress-maker on a courtly assembly of individuals
distinguished by rank and talent. The same very
common error, arising from the weakness or short-
sightedness of ordinary human nature, of confound-
ing means with ends, is strikingly exemplified in the
case of the person, who perceiving that money
will procure all desirable things, at last becomes
the insane miser, and dies from want of the common necessaries of life rather than touch his
hoarded treasures :—it is seen also in the case of

the mere pedagogue, who values the abstractions
of language or grammar and of mathematics more
than the useful knowledge which they may serve
to convey; or, lastly, still more suitably for our
purpose, in the case of the bibliomaniac, who regards the type and binding of his books more than
the subject. To prove how unessential what is
called high-finishing in painting, is to the complete
obtainment of the purposes of the art, we have
the cartoons of the immortal Raphael, which to
the mere mechanic appear almost to be daubs :
and many of the mere sketches of genius are to
a true taste more precious than some of the most
finished pieces in our galleries. Again, of what consequence is it to a man whether he see approaching
the friend of his heart by daylight or candlelight,
or with the source of light above or below, &c.,
provided there is light enough for him to distinguish clearly the friend of his heart. — A painter
will discover the difficulties which a brother artist
had to surmount in representing an object in some
particular predicament, as regards the light, &c.,
and may estimate the talent accordingly ; but the
great mass, even of the most accomplished ordinary spectators, will generally be looking beyond
the sign to the thing signified ; and perhaps few
will at all heed the difficulties. In consequence of
the prejudice in favour of "a sweet or adorable bit
of colouring," as it will sometimes be called, and
which in truth may have the merit of most natural
colouring, there are preserved in many galleries
pictures disgusting in almost all other respects,—as
of drunken Dutch boors, with fiery noses and phy-

siognomies degrading to human nature, &c. &c. ;
on seeing which the man of taste deplores that the
art of representing should be prostituted to the vile
purpose of representing things of worse than no
interest, and that its nobler objects should be so
often forgotten.

" *When the image formed, as above described, beyond a lens, is viewed in the air by an eye placed
still farther beyond in the same direction, the arrangement, according to minor circumstances, constitutes either the common* TELESCOPE *or* MICROS-
COPE." (Read the whole second paragraph of
the Analysis, page 162.)

The name of *telescope* (a compound Greek term,
signifying to *see far*, as *microscope* signifies to *see
what is small*), applies to that wondrous instrument
of modern invention by which the intelligent soul
on the beams of light as its path, is enabled
as it were, to dart widely into space for the
purpose of contemplating the distant glories of
creation ; or, again, by which distant objects
are instantly brought near to the eye for the
purpose of convenient inspection. In ancient
times, a man, while looking with admiration on the
bright face of the moon, might have exclaimed,
" Would that I had the power to fly upwards to
that celestial orb, the better to understand its nature and beauties ;" but he could little have
dreamed that the day was coming when human
ingenuity would in effect achieve the wish :—now
by the telescope it is atchieved, for an instrument
which merely doubles apparent magnitudes, shews

the moon exactly as she would appear to a person who had ascended towards her from the earth a distance of 120,000 miles, while one of greater power produces effects correspondingly great. But to examine the heavenly bodies is by no means the only use of the telescope, man being often more concerned to discover what is passing on the surface of the earth around him. Thus by a telescope, the military chief descries approaching friends or foes, while yet concealed from others and from the naked eye, in the blue mist of distant mountain or plain—or similarly, the sailor, whose attention has been attracted to a little speck on the sea-horizon, discovers there a ship of class and nation at once evident to him, and with the crew of which, by the additional use of signal flags, he is enabled readily to converse :—at midnight a telescope directed to a distant cathedral, will so effectually call it into the presence of the observer, that on the clock-turret may be read from the slow moving hands the unceasing lapse of time. A man, in the midst of a wide plain, or on a lofty hill-top, or far on the face of a lake, nay, even in his own garden, or in his house near some open window, who might suppose himself quite alone and unseen, would yet by a telescope be instantly placed under the observation of whoever chose to watch him. Some remarkable cases of actions, imagined by the parties to have been done in perfect secrecy, have thus been brought to light.

Now the telescope with its extraordinary powers exhibits but another modification of the simple case described at page 194, and exemplified in

the camera obscura, &c., of an image formed for
visual inspection beyond a lens. And we shall
here explain that its powers depend altogether on
the two circumstances, first, of its large lens collecting for the formation of the image (subse-
quently transferred to the observer's retina) a
thousand or more times the quantity of light which
the naked pupil could receive; and, second, of its
forming by this light an image, which to the eye
brought near it appears vastly larger than the object itself appears.

To understand this well, we must recall, that
the nature of the bending of light in passing
through a lens is such, that all the rays reaching
the lens from any point of a visible object in front,

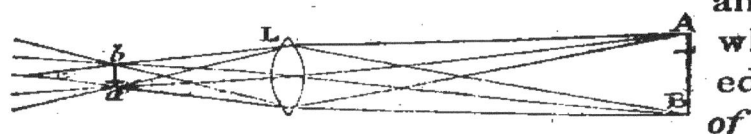

and forming
what is call-
ed a *pencil
of light*,—as
that spreading from the point A of the cross here
represented to the lens L—are collected in a corresponding point, as *a*, at the focal distance beyond
the lens, and so as always to meet the central ray
of the pencil; and therefore when the light comes
from above the centre of the lens, the focal meeting is below, as shewn here; and when it comes
from below, the meeting is above: then the same
happening as regards every visible point of the object (the rays from only the two extreme points A
and B are here represented) at corresponding points
beyond the lens in the space between *a* and *b*, the
collected light, if received on a white screen
placed there, as in the camera obscura, will make

apparent to an eye in any direction a beautiful
inverted image of the object. Now in the place
where the rays meet to form this image, if no
screen be interposed, the rays are not lost or
destroyed, but merely cross each other in the air
nearly as they previously crossed, without inter-
ference, in the lens, and spread again beyond the
focal points, or towards *c*, as here shewn, as they
originally spread from the several points of the
object itself: an eye therefore placed any where
beyond *c*, must receive portions of the pencils
from every point of the image, and may see it in
the air as an object situated in the focus of the
lens.—This may be observed at once by holding a
spectacle glass or any lens at proper distance between an object and the eye.

Now a telescope is merely a long tube, blackened within to exclude and destroy useless light,
and having a large lens called the *object-glass*,
filling one end of it as a window, to gather the
light from the objects in front, and to form with
it images near the other end of the tube, where
they may be conveniently inspected. These images,
for a purpose to be immediately explained, are
examined through another lens called the *eye-
glass*, which is fixed in a smaller tube made to
slide backwards and forwards in the larger, so as
to admit of the focal distances being adjusted to
the power of different eyes, &c. The accompanying sketch
shews the
progress of
the light

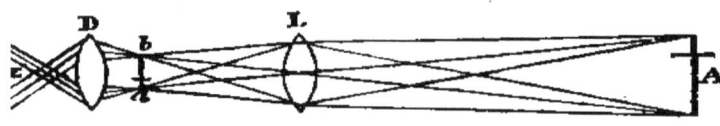

from the object A, through the object glass L, to form an image at *b a*, and afterwards to be bent by the eye-glass D, so as to enter the pupil of the eye at E, to form the last image on the retina.

In the simple telescope with only two lenses, as above represented, called the *astronomical teles-cope*, the image is inverted : but this is a circum-stance of no importance in viewing the heavenly bodies ; to fit the instrument, however, for view-ing terrestrial objects, it is necessary to place in the tube another simple or compound lens, which shall form a second image from the first, and by inverting a second time, shall produce an image really upright.

To determine how much larger an object will appear when viewed through a certain telescope, for instance, one with an object glass of three feet focus, than when viewed by the naked eye, we must recollect that the image is formed in the focus of the object glass, or at *b a*, and subtends from the centre of that lens, as at *c*, the same visual angle as the object itself (a fact explained at page 204), and to an eye placed there, would appear of the same size as the object ; but if the eye can be brought fifty times nearer to the image, this will appear fifty times taller and broader, and therefore with 2,500 times the surface, and thus, as compared with the object, may be called much magnified. Now as the naked eye cannot see distinctly an object nearer to it than at about six inches, because of the great divergence of light from a nearer radiant point, the telescope in ques-tion, without an eye-glass, would allow the eye to come only six times nearer to the image than when at the centre of the object-glass, and would only magnify the diameter six times ; but if then an eye-glass, as D, of half an inch focus, were placed half an inch from the image, so as to render the rays of every pencil parallel, and therefore fitted to the powers of the eye, while the different parcels would cross each other a little way beyond the glass, as shewn above, an eye placed to receive in its pupil the crossing parcels, would see the image as large as if at half an inch from it, and therefore 72 times nearer than if viewed from the object-glass, and therefore again as of 72 times greater diameter.   Now, as in all cases, the image in a telescope is in the focus both of the object-glass and eye-glass, and is therefore nearer to the latter than to the former in proportion as their focal distances differ, the magnifying power is measured by that difference—in the case at present supposed the difference is as 72 to 1, and 72 is the magnifying power of the telescope.   The rule is generally thus expressed, " divide the focal distance of the object-glass by that of the eye-glass, and the quotient is the magnifying power." It is always to be remembered, that if the diameter of an object be magnified ten times, the surface is magnified 100 times, and so in proportion for other numbers.

With such means of aiding the sight, then, is it that we discover the mountains of our moon, and can even measure their altitudes ; that we can see the four beautiful moons of the planet Jupiter ; that we can perceive marks and irregularities on the surfaces of the other planets, enabling us to say at what rate they severally whirl round their axes, experiencing the phenomena of day and night;—and that we can determine many other interesting particulars.

The discovery of the telescope is said to have been first made accidentally by the children of a Dutch spectacle-maker, while they were playing with their father's work; but it was turned to no use until Galileo, led by science, fell upon it again, and with the knowledge of its worth, obtained from it the most sublime results. If ever human heart throbbed with delight, it must have been when Galileo first directed his optic tube to the heavens, and through it contemplated so many glorious objects which no human eye before had seen!—as Venus, our beautiful morning and evening star, appearing not a circle, but a crescent like our moon in her quarters—as the satellites of Jupiter—the rings of Saturn—myriads of stars until then invisible to man; and in a word, when he beheld the undoubted proofs of the true system of the universe, as his genius had before conceived it, uniting the greatest simplicity with its unspeakable grandeur.

The Galilean telescope was simply a large object-glass to collect much light, with a small concave eye-glass placed so as to intercept the converging rays before they reached their focus, and to change their convergency into the parallelism which the eye could command. This telescope, although magnifying less than that made of two convex glasses, as above described, still from occasioning no loss of light by the crossing of rays in forming an image, was of considerable power. The same principle is now adopted for the common opera glass.

It was explained at page 192, that a ray of light, in being bent or refracted by transparent media, as by a lens, is also divided into rays of the different colours seen in the rainbow. Hence an image formed behind a simple lens has coloured edges or fringes. This fact rendered the images of small objects, formed in the first telescopes, very indistinct; and, but for the important discovery originally made by Dollond the optician, that different kinds of glass have the *dispersive* and *refractive* powers with different relations, so that a concave lens of a certain curve applied to a convex lens might completely counteract the dispersion of colour by the latter, while it left enough of the convergence of the rays for the formation of an image—refracting telescopes would have always been very imperfect. Dollond called his telescopes *achromatic, or without colouring power.* It is very remarkable, that he had the fortune to obtain some glass for his purposes more suitable than any which has been procured since, or which could be made by known rules, until the late improvements in the manufacture suggested by the ingenuity of Mr. Farraday. The author carried abroad with him a small Galilean telescope of Dollond's, which often gave more correct information respecting minute coloured objects at a distance, as signal flags at sea, than much larger glasses of modern make.

The MICROSCOPE, of greatest power and with

the form called compound, in its structure approaches very closely to the *telescope*, the chief difference being, that while in the telescope a large distant object forms in the focus of the object-glass an image exactly as much smaller than itself as the distance of the image from the glass is less,—in the microscope conversely, a small object placed near the focus of the object-glass produces a more distant image, as much larger than itself as the image is more distant,—and in the one case as in the other, the image is viewed through an appropriate eye-glass. The object-glass in the telescope is large, in the microscope it is generally very small. If, in the latter, an object-glass be used of one-eighth of an inch focal distance, and the object be so placed that its image is formed at six inches, the image will be of diameter 48 times as great as the object, or will have nearly 2,500 times as much surface; and if that image be viewed through an eye-glass of half an inch focus, the image will appear still twelve times larger, or 30,000 times larger than the object.

One convex lens is called a single microscope, and it magnifies, as already explained, chiefly by allowing the eye to be brought so much nearer to the object than would be possible to see it without the glass : but even if the distance of the eye and object is not changed, a lens interposed will still

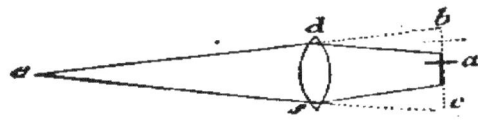

magnify by bending the light, as at *d,* and making that which comes to the eye at *e*

from the top of such an object, as the little cross
a, appear to come from *b*, and that from the bottom to come from *c*, thus magnifying the cross here
represented by the black lines, to appear of the size
represented by the dotted lines. A concave lens
minifies for the contrary reason.

Perhaps there is not a greater treat for a person
who has feeling for the beauties of nature, than to
explore with the microscope. While the telescope
lifts the mind to the contemplation of boundless
space, occupied by myriads of suns, and exhibits
this globe of ours as less, compared with the universe around it, than a leaf is compared with a
forest, or a grain of sand compared with all which
lies on ocean's shore; the microscope, again,
excites new astonishment by shewing on a leaf, or
in a single drop of water in which the leaf has
been infused, thousands of living creatures, and
of creatures not imperfect because thus small, but
endowed with organs and parts as complex and
curious as those of an elephant. And he who
admires the curious structure of a honey-comb,
may bend his eye through the microscope upon
the cut surface of a willow-branch, or of other
wood, there to see a similar structure more wonderful still : or he may compare the lace of a fly's
wing with the most perfect which human art can
weave ; or the beautiful proportions and perfection
of the limbs and weapons of an insect, invisible
perhaps to the naked eye, with any larger objects
of the kind already known to him.

Telescopes and microscopes might with propriety be both called microscopes, for often the

telescopic object appears to the naked eye even smaller than that which the microscope examines. The minutest visible insect at hand may hide from the eye a planet at a distance. The image in the telescope is smaller than in the microscope, because the rays from a distance being nearly parallel, must form the image nearly in the focus of the object-glass; while for the microscope, the rays from the near object being very divergent, form the image far beyond the focus, and proportionately larger.

" *Light falling on very smooth or polished and flat surfaces, is reflected so nearly in the order in which it falls, as to appear to the eye as if coming directly from the objects originally emitting it,— and such surfaces are called* MIRRORS." (Read the Analysis, page 162.)

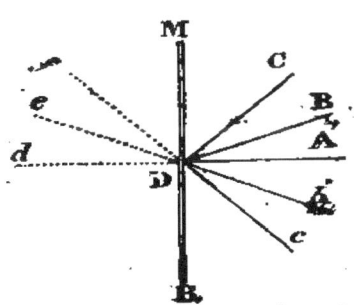

If a marble slab, or other flat surface, were in the situation M R, with its edges towards the spectator, a ball projected from A perpendicularly towards it at D, would rebound directly back to A, but if projected from it obliquely, as from B to D, it would not return to the first situation B, but to *b,* a situation as distant on the opposite side of the perpendicular, thus making the *angle* of the return or *reflection* equal to the *angle* of approach or *incidence ;* the same would be true of a ball approaching obliquely from any other point, as C, and returning to *c.*

Now light is reflected from polished surfaces according to the same law, so that an eye at A would see itself as if placed at *d*, an eye at *b* would see an object really at B as if it were at *e*, and so forth. Where the existence of a mirror is not suspected, the objects reflected from it are held to be realities placed beyond where it is. A wild animal will attack its image in a glass; and a dog crossing a brook, will quit the piece of meat in its mouth to catch the tempting image which the eye sees in the water below. The reason that an object seen in a plane mirror appears to be just as far beyond the mirror as its true distance on the side of the spectator, is, that the diverging rays of a pencil of light have the same divergence after as before reflexion.

Any plane very smooth surface reflects light as now described, and is a mirror; but different substances send back very different proportions of the light which falls on them. A highly polished metallic surface is the best mirror, often returning three-fourths of the whole light. Hence in reflecting telescopes, the mirrors are made of polished metal.

Our common looking-glasses are really metallic mirrors, for it is the smooth clear surface of the quicksilvered tin foil behind the glass which reflects the light, the glass itself merely serving the purpose of preserving the metallic surface perfectly clear and flat. There is always an imperfection in such glass mirrors, when used for viewing oblique objects, because the glass bends the light a little, and because the external surface of the glass acting also as a mirror, although so much more feebly than the metal behind, forms an image which mixes with and confuses the other.

The mirror-power of glass unaided is seen from the panes of a plate-glass window, which make objects in front very visible, although by no means with clearness comparable to that from a metallic surface. All common panes of glass in windows, or covering print-frames, &c., reflect as much light as plate-glass, but the reflection being irregular because the surface is irregular, it does not attract notice.

The smooth surface of a fluid is a mirror, and which moreover is horizontal; and when that surface is metallic, as of mercury, the mirror is most perfect. In water, spirits, oil, or any other liquid, it is also perfect, but feebler.

The mirror of liquid quicksilver is sometimes used by astronomers in observing the apparent altitudes of the heavenly bodies, for the image in the mirror appearing exactly as much below the horizon as the object is really above it, half the distance between them is the true height.

A varnished picture, or any japanned surface, is a mirror; nay, also, even a polished piece of wood for instance a mahogany table,—as is well known among playful children. The author, while writing this, is looking on a table covered with black leather, and in that covering, as a mirror, he clearly sees all bright objects beyond the table. Polished stones, as marble slabs, &c., reflect as much as glass. But even a surface of air may be a mirror, as where a cold and dense stratum happens to be in contact with a warmer and rarer stratum,—hence the trees, islands, &c. reflected from below, and seen in the sky where particular causes have unequally heated different levels of the atmosphere. On the burning sands of Africa a stratum of air near the surface being heated, that above sometimes becomes a mirror, thus reflecting more remote objects : certain kinds of mist and thin clouds will elsewhere produce a similar effect, so that in them a ship may be seen as if suspended aloft, with keel uppermost.

An object seen by the light reflected from a mirror appears always reversed, as for instance, when the right hand of a person standing before a glass becomes the type for the left hand of the image—and it may be a stump : or when a tree or rock, or mountain, seen in the mirror of a lake, has its top downwards.

It is on this account, that a man painting his own portrait from a mirror, is apt to reverse all accidental characteristics of the countenance or person, unless they are the same on both sides ; and then if, as is generally true, one eye be higher than the other, or the nose be a little to one side, a very incorrect resemblance will be produced. Hence a person with a countenance at all thus peculiar, never sees himself in a mirror as he appears to others; and a belle or beau, who has decided that a curl is more graceful on the left temple, may unconsciously leave it on the right.

By reflecting any image, however, from a first mirror to a second, and from that to the eye, persons may see the object, or themselves, if they choose, as others see them. What a pity that there are not some moral mirrors to answer an analogous purpose, and occasionally to tell persons how hideous they are to virtue's eye, although they think of themselves with such complacency!

A candle placed between two parallel mirrors, makes visible in either glass to a spectator on one side an endless straight line of lights. If the glasses be inclined to each other, the lights will appear as if in the circumference of a circle, having its centre where the prolonged mirrors would meet : this fact is well illustrated in the beautiful toy called the *kaleidoscope*. By placing several mirrors in particular situations around an apartment, a man entering it may see himself multiplied into a crowd, and a few ornamental pillars may produce the effect of thousands formed into long colonnades or retiring lines.

The sun or moon reflected in a still lake, appear as they do in the sky ; but if the surface of the water become at all ruffled by the breeze, instead of one distinct image, there will be a long line of bright tremulous reflection. The reason of this appearance is, that every little wave, in an extent perhaps of miles, has some part of its rounded surface with the direction or obliquity which, according to the required relation of the angles of incidence and reflection, fits it to reflect the light to the eye, and hence every wave in that extent sends its momentary gleam, which is succeeded by others.

Although the external surface of glass does not reflect but a small part of the light which falls upon it, being therefore a feeble mirror, very curiously, if light, which has entered a piece of glass, fall upon a back or internal surface, very obliquely, instead of passing out there, it is more perfectly reflected than it would be by the best

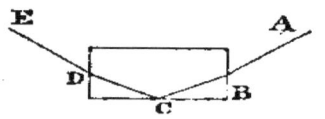

metallic mirror. Thus light from A entering at B, is entirely reflected at C,? and escapes at D towards E. The back of a wedge of glass, or common prism, thus becomes a perfect mirror.

It is this fact which enabled Dr. Wollaston to construct that beautiful little instrument called by him the *Camera Lucida*. The two surfaces at the back of the small prism of glass A become mirrors, the first reflecting to the second, and the second to the eye at E, the objects in the landscape before it while the eye also sees through the glass to the paper below at B, and may suppose the imagery to be feebly pourtrayed on the paper; with a pencil that appearance is made permanent, and a correctly drawn outline of the scene is at once obtained. This instrument for assisting draughtsmen is still simpler than the camera obscura. Other modifications of the instrument have since been contrived.

The same fact of the back and internal surface of a transparent mass becoming a mirror, gives us the explanation of that phenomenon so admired before it was understood, and not less admired since, *viz.* the *rainbow*, or *arc in the sky*, as in France and elsewhere it is named—the object which the poets of nature have almost worshipped for its beauty, one of the delights of our boyish days, when we saw it stretching over the haunts of our young pleasures, and when we pursued it in the hope of catching some of the falling rubies and emeralds, or bright-coloured dew of which it might be composed.

When a partial shower of rain falls on the side of the landscape opposite to where the sun is shining, this beautiful phenomenon immediately appears, *viz.* a variegated arch, red at its external border or confine, and then successively orange, yellow, green, &c., (in the order of the colours of the prismatic spectrum described at page 190,) towards its inner border or confine: its centre is directly opposite to the sun, as if at the end of a straight line drawn from the sun through the eye of the spectator towards the opposite horizon, and being therefore always under the horizon, the bow is less than a semicircle. The diameter of the circle of which the bow is a part is of about 82° of the field of view. There is a second bow of much fainter light than the first, and with the colours in reverse order: it is of 108° diameter, and therefore external to the other.

Now the explanation of this miracle of beauty is simply as follows. While the sun shines upon the spherical drops of falling rain, and its light falling upon the whole central part of any drop, passes completely through, still that portion which enters near the edge of the drop, as at *a*, is refracted, and reaches the back surface of the drop at

x

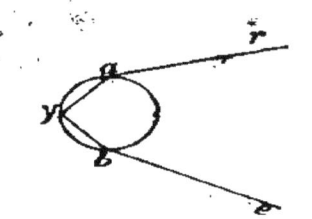

y so slantingly, or at an angle so great, that there occurs an entire reflection of it instead of transmission; the ray therefore is returned to *b*, where it again escapes from the drop, and as here shewn, descends to the earth or eye at *e*. Thus every drop of rain on which the sun shines is a little mirror suspended in the sky, and is returning at a certain angle all round it, *viz.* at an angle of 41°, a portion of the light which falls on it; and an eye placed in the required direction, receives that reflected light. If in this case there were *reflection* only, and not also *refraction* with *separation of colours*, the rainbow would be only a very narrow resplendent arc of white light, built up of millions of little images of the sun; but in truth, because the light which enters near the edge of the drop traverses the surface very obliquely, it is much bent or refracted before its reflection, as seen at *a*, and is divided into rays of its seven colours, as it would be on passing through a prism (as explained at page 192); and this division or separation continuing after the light again escapes from the drop at *b*, instead of one white ray descending from each drop to a certain point of the

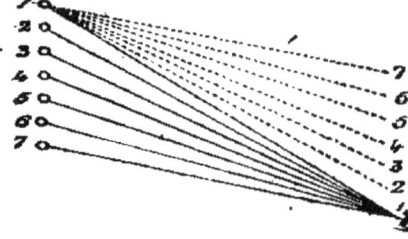

earth, seven rays descend (here marked by dotted lines from the figure 1 on the left hand, to 7, 6, 5, &c. on the right), and of these an eye can only

catch one at a time : but for the same reason
that seven eyes placed in a line from above downwards (viewing the centre of the bow) would be
required to see the seven colours from one drop,
so one eye looking in the direction of seven drops
situated in a corresponding line, as from 1 to 7,
will catch the lower or red ray of the upper, the
orange or second ray of the next, the yellow or
third ray of that which follows, and so on, while
it will lose all the others, and thus will see the several drops as if they were each of one colour only.
Of such elements, then, found in the same relative
directions all around the eye, the glorious arch is
formed. No two eyes can see the same rainbow,
that is, can receive light from the same drops at
the same time ; and the same eye does not for two
instants receive light from the same drops. This
rainbow can never appear to a person on a plain,
unless when the sun is within 41° of the horizon,
for otherwise the centre of the rainbow would be
more than 41° under the horizon, and 41° is the
whole semidiameter of the bow.

· We have described above what is called the principal bow, formed in the drops by two refractions,
and one reflection of light. To produce the
fainter second or external bow, mentioned above,
and of which the colours are in reverse order,
the light which enters at *a* is reflected first at *y*,

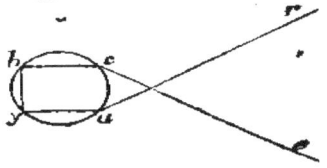

then again at *b*, and escapes at *c* towards the eye :
hence there are two reflections as well as two re-
fractions. As the semi-

diameter of this bow is 54°, it may be visible when the internal bow is not.

An artificial rainbow may be produced in sunshine at any time by scattering water-drops from a brush or otherwise. The cut-glass ornaments of chandeliers, &c. produce colours on the same principle as rain-drops; as do also mist and particles of frozen water between a luminous body and the eye, exhibiting the circular *halos* often observed around the sun and moon.

" *Mirrors may be plane, convex, or concave; and certain curvatures will produce images by reflection, just as lenses produce images by refraction; so that there are reflecting telescopes, microscopes, &c., as there are refracting instruments of the same names.*" (See the Analysis, page 162.)

While a plane surface reflects light, so that what is called the image in it of a known object may readily be mistaken for the reality, convex or concave mirrors reflect as if every distinct point of them were a separate small plane mirror, and their effects on light correspond with the relative inclination of the different parts. The only forms of much importance are the regularly spherical or parabolic concave and convex mirrors. We shall now find that these have on light similar effects with lenses, only the concave mirror answers to the convex lens, and the convex mirror to the concave lens. It is the concave mirror which gathers the light to form images in the most perfect telescopes that exist, as those of Herschell and others. Admirable as in certain respects is the refracting telescope, it falls much short of the telescope acting by reflection.

In a hollow sphere, or part of a sphere with polished internal surface, if rays radiate from the centre in all directions, they reach every part perpendicularly, and therefore are thrown back to 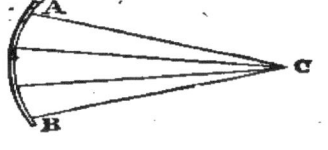 the centre. Thus, if A B were a concave spherical mirror, of which C were the centre, rays issuing from C would again meet there.

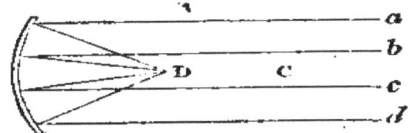 It can be proved also, that any ray parallel to the axis, falling upon such a mirror, will be reflected inwards so as to cut the axis half-way between the mirror and its centre, *viz.* at D. Then as all parallel rays meet in the same point, that point becomes a focus, as already explained for lenses;—about that part an image of the sun, for instance, will be formed when the mirror is held directly towards the sun. This point is called the principal focus of the mirror.

For the same reason that parallel rays meet in the focus, so will rays, issuing from the focus, become parallel, after reflection, as seen in the figure at page 59 ; and if they be then caught in a second and opposite mirror, as there also represented, corresponding effects will follow.

Now, for a concave mirror, as already explained for a lens, when rays fall on it obliquely from one side of the axis, their focus will be on the opposite side, and therefore the mirror will form an in-

verted image of any body placed before it, just as
a lens does; and the image will be near or distant, and large or small, according to the divergence of the approaching rays, exactly as happens
with lenses: and thus, the camera obscura, magic
lantern, telescopes and microscopes, may all be
formed by mirrors, as they may be by lenses.
Moreover, concave mirrors magnify, and convex
mirrors minify, as concave lenses of the opposite
names do. The two subjects of images by refraction and by reflection run so nearly parallel, that it
would be useless repetition here to enter upon the
detailed consideration of the latter subject, 'and
we shall therefore content ourselves with shewing
why a concave mirror magnifies, and why a convex
mirror minifies.

A concave mirror
magnifies, because the
light from A reaching
the mirror where it
can be reflected to an eye placed at F, *viz.* at E,
seems to the eye to come from C, and the light of
B similarly appears to come from D, so that the
cross A B, by the reflection, seems to the eye to be
of the greater dimensions C D.

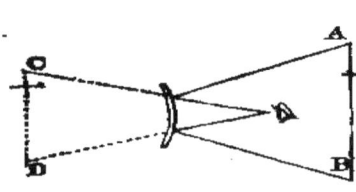

In a convex mirror, again,
for corresponding reasons, the
cross A B appears only as
C D, and therefore much
smaller than the reality.

Concave, or magnifying mirrors, are often used
by persons in shaving.
A convex mirror is a common ornament of our

apartments, exhibiting a pleasing miniature of the room and its contents.

Any polished convex body is a mirror, and therefore the ball of the human eye is one in which we may contemplate most perfect miniatures of surrounding things. It is the image of the window, or of the sun, in the convex mirror of the eye, which painters usually represent by a spot of white paint there; and a similar luminous spot or line must be made when they represent almost any of the pieces of furniture which have rounded polished surfaces, as bottles, glasses, smooth pillars, &c.

Convex lenses thus are mirrors to all the objects around them, and very strikingly so, owing to the perfection of the form of a lens. The polished back of a watch, often in the same way attracts the attention of a child, who wonders to see there so clearly ' a little baby.'

It has been a mathematical amusement to calculate what kind of distortion mirrors of unusual forms will produce, and then to make distorted drawings, which reflected from such mirrors, might produce in the eye the natural image of the objects.

When a concave mirror is used for a telescope, the image formed in front of it, and examined usually through a powerful magnifying eye-glass, may be viewed,—as in Herschell's telescope, by the spectator turning his back to the real object, and looking in at the mouth of the telescopic tube, near to the edge of which the image is thrown by a slight inclination of the mirror at its bottom :—

or as in the *Newtonian* telescope, through an opening in the side of the tube, after being reflected by a small plane mirror placed diagonally in the centre of the tube:——or as in the *Gregorian* telescope, through an opening cut in the principal mirror or speculum, after being reflected towards that opening by a smaller mirror placed in the centre of the tube : this last arrangement is that preferred for smaller telescopes, because the spectator, while seeing the image, is also looking in the direction of the object.

Reflecting telescopes have the advantage of being perfectly *achromatic*, that is, of producing no coloured or rainbow edges to the images : for compound light is reflected, although not refracted, entire, all the colours following the same law of equal angles of incidence and reflection.

Herschell's largest telescope had a mirror of 48 inches in diameter, and therefore received about 150,000 times more light than an unassisted eye could, making with it, at a focal distance of 40 feet, a large image admirably distinct. It was with this that, in the obscurity of remote space, he discovered rolling along the immense planet, which in honour of his royal patron, he called the *Georgium Sidus*, but which now, by the decision of the, scientific world, bears his own name;——and with this he discovered moons before unseen, of other planets, and he unravelled the celestial nebulæ and clustered stars of the milky way, and, in a word, unveiled, vastly more than had before been done, the system of the boundless universe. If this world were to last for millions of years, the discoveries of Herschell's telescope would mark a memorable epoch of its early history.

" *Light returned from, or passing through bodies of rougher or irregular surface, or which have other peculiarities, is so modified as to produce all those phenomena of colour and varied brightness .seen among natural bodies, and giving them their distinctive characters and beauty.*" (See the Analysis, page 162.)

General remarks on this part of our subject were made in the beginning of the section, in the explanations of how objects not self-luminous become visible by reflecting the light of other bodies, and of how the prism separates a ray of white light into rays of the several colours of the rainbow—which rays, on being again mixed, reproduce white light as before :—and much beyond these remarks we have not the intention of now proceeding. To give a full account of the matters that come within the scope of this department, would occupy the pages of a large volume, for there would be to pass in review the various opinions which have existed on the *intimate nature* of light,—the facts connected with what has been called the *polarisation* of light,—the relation of light in *double refraction*, to the ultimate structure of material masses, &c., all which subjects are in certain respects highly interesting, but —as some of them are not yet completely investigated—as respecting others, various opinions prevail,—as they involve few matters applicable to common use,—as the reasonings about them are far removed from ordinary trains of thinking, and refer to facts altogether unknown to common observation,—we hold them not to be fit parts of a popular treatise on light. We may state, however, that persons who have the leisure and the mathematical preparation necessary for pursuing the study, will find their labour in it richly rewarded.

What we deem it necessary then here to add is, that white light in falling upon any transparent substance, as air, water, glass, &c., reduced to thin plates or films, is so affected, that for certain degrees of thinness, different for each substance, it is decomposed, and is reflected or transmitted not as white light, but as some of the colours of the rainbow, and the colour reflected in any case is always the opposite or complement of that which passes through, that is to say, such that the two brought together again make white light. The facts may be studied, as Newton originally studied them, in the thin plate of air which occupies the space between a convex lens and a plane surface of glass upon which the lens is laid,—in which plate, as the distance from the point of the apparent contact of the glasses increases, there appear successive rings of vivid colours. The same truth is exemplified, in the colours of a soap-bubble, which brighten as the bubble swells and becomes of thinner substance, and are different as the thickness increases from above downwards;—and it is exemplified in numerous other common facts. Now whatever be the reasons of such decomposition of light—

and the explanation is not yet complete—we cannot doubt that in natural bodies generally, the colours, opacity, transparency, &c. depend entirely upon the volume and arrangement of the minute fibres or plates, with included interstices, which constitute the volume or structure of each mass. Accordingly, whatever changes that arrangement, may change also the colour of the mass. Thus by drawing a certain number of minute lines on a certain extent of any metallic surface, we may make it of what colour we please; and mother-of-pearl owes its beauty entirely to its furrowed or striated surface, as is proved by our taking an impression of that surface on sealing-wax, and perceiving that the wax then exhibits similar colours.

The investigations in progress respecting the phenomena of light, are furnishing new proofs of the extreme simplicity of nature, amidst the boundless extent and infinite variety. When men thought of the sense of touch only as it exists at the tips of the fingers, or on the general surface of the body, they were far from suspecting that the sense of hearing had the near relation to it which subsequent discoveries have proved, and still less did they think, that the sense of sight was similarly related; but step by step they ascertained, 1st. of sound coming to the ear through the air—that air was a material fluid as much as water, consisting of the same or similar particles, only more distant among themselves—that a motion or trembling in the air, by affecting nerves exposed in the ear, produced the sensation of sound, as the trembling in a log of wood caused by the action of a saw produces a peculiar sensation of touch in a hand laid on the log,—and, finally, that common sound in all its varieties, is merely such trembling of the air, affecting a structure of nerve so exposed in the ear, as to be as much more readily excitable than the nerves in the fingers and elsewhere in the skin, as the action or impulse of moving air is more delicate than that of common solids and liquids. And now, in the investigations respecting light, this kind of comparison is carried a step further, for it is become matter almost of certainty that the sensation of light is produced in a suitable nervous tissue in the eye, by a trembling motion in another fluid than air, which fluid pervades all space, and in rarity or subtlety of nature surpasses air vastly more than air does water or solids;—and while in sound, different tones or notes depend on the *number* of vibrations in a given time, so in light do different colours depend on the *extent* of the single vibrations. Can human imagination picture to itself a simplicity more magnificent and fruitful of marvellous beauty and utility than this!—But farther, as air answers in the universe so many important purposes besides that of conveying sounds, although this alone comprehends language, which almost means reason and civilization—so also does the material of light minister in numerous ways, in the phenomena of heat, electricity, and magnetism.

The truths now positively ascertained with respect to the nature of light and vision, are perhaps those in the wide field of human inquiry which, acting on ordinary apprehension, most forcibly place the individual as it were in the presence of Creative Intelligence, and awaken the most elevated thoughts of which the human mind is capable. Had there been no light in the universe, all its other perfections had existed in vain. Men placed on earth would have been as human exiles with their eyes put out, abandoned on an unknown shore, of climate and productions totally new to them : every movement might be to destruction, for their perceptions would be limited by the length of their arms, and of their fearful groping steps, and the wretched beings, separating when impelled by hunger to search for food, would probably scatter to meet no more. But the material of light exists, pervading all space, and certain impressions made upon it in one place rapidly spread over the universe, the progressive impression being called a ray, or *beam of light.* The beams of light, then, from all parts coming to every individual, may be regarded as supplementary arms or feelers belonging to the individual, and which reach to the end of the universe, so that each person, instead of being as a blind point in space, becomes nearly omnipresent : — then these limbs or feelers have no weight, they are never in the way, they impede nothing, and they are only known to exist when their use is required ! But this miracle of light would have been totally useless, and the lovely paradise of earth would have been to man still a dark and dreary desert, had there not been the twin miracle of an organ of commensurate delicacy to perceive the light, *viz.* of the *Eye* ;—in which there is the round cornea of such perfect transparence, placed exactly in the anterior centre of the ball (and elsewhere it had been useless), then exactly behind this, the beautiful curtain the iris, with its pupil dilating and contracting to suit the intensity of light—and exactly behind this again, the crystalline lens, having many qualities which only complex structure in human art can attain, and by the entering light forming on the retina beautiful pictures or images of the objects in front,—the most sensible part of the retina being where the images fall. Of these parts and conditions, had any one been otherwise than as it is, the whole eye had been useless, and light useless, and the great universe useless to man, for he could not have existed in it. Then, farther, we find that the precious organ the eye is placed not as if by accident, somewhere near the centre of the person, but aloft on a proud eminence, where it becomes the glorious watch-tower of the soul ; and, again, not so that to alter its direction the whole person must turn, but in the head, which on a pivot of admirable structure moves while the body is at rest ; the ball of the eye, moreover, being furnished with muscles which, as the will directs, turn it with the rapidity of lightning to sweep round the horizon or take in the whole heavenly concave ;—then is the delicate orb secured in a strong socket of bone, and there is over this the arched eyebrow as a cushion to destroy the shock of blows, and with its inclined hairs to turn aside the descending perspiration which might incommode ;—then is there the soft and pliant eyelid with its beauteous fringes, incessantly wiping the polished surface, and spreading over it the pure moisture poured out by the lacrymal glands above, of which moisture the superfluity by a fine mechanism is sent into the nose, there to be evaporated by the current of the breath :—still further, instead of there being only one so precious organ, there are two, lest one, by accident, should be destroyed, but which two have so entire a sympathy, that they act together as only one more perfect ;—then the sense of sight continues perfect during the period of growth from birth to maturity, although the distance from the lens to the retina is constantly varying;—and the pure liquid which fills the eye, if rendered turbid by disease or accident, is by the actions of life, although its source be the thick red blood, gradually restored to transparency. The mind which can suppose or admit that within any limits of time, even a single such organ of vision could have been produced by accident, or without design,—and still more, that the millions which now exist on earth, all equally perfect, can have sprung from accident—or that the millions of millions in past ages were all but accidents—and that the endless millions throughout the animate creation, where each requires a most peculiar fitness to the nature and circumstances of the animal, can be accident—must surely be of extraordinary character, or must have received unhappy bias in its education.

As a concluding reflection with respect to vision

we may remark, that all the provisions above considered have mere utility in view, for any one of
them wanting would leave a necessary link in the
chain of creation wanting : but, we have shewn
in a preceding part of the work, that if there had
been white light only, susceptible of different degrees of intensity and shade, the merely useful purposes of vision would have been answered about
as perfectly as with all the colours of the rainbow
—which truth is instanced in the facts, that many
persons do not distinguish colours, and that it im-
ports not whether a person view objects in the
morning, or at mid-day, or at even-tide, or through
plane glass or coloured glass.   While, therefore,
the existence of light generally, and of the eye,
speaks of Creative Power and Intelligence, the existence of colours, or of that lovely variety of hues
exhibited in flowers, in the plumage of birds, in
the endless aspects of the earth and heavens ; in a
word, in the whole resplendent clothing of nature,
—because appearing expressly planned, as a source
of delight to animated beings, speaks of Creative
Benevolence, and may well excite in us towards
the Being in whom these attributes concentrate,
the feelings associated in our minds during this
earthly scene, with the endearing appellation of
' Father.'

END OF THE SECTION ON LIGHT.

LONDON:
PRINTED BY J. L. COX, GREAT QUEEN STREET,
Lincoln's-Inn Fields.

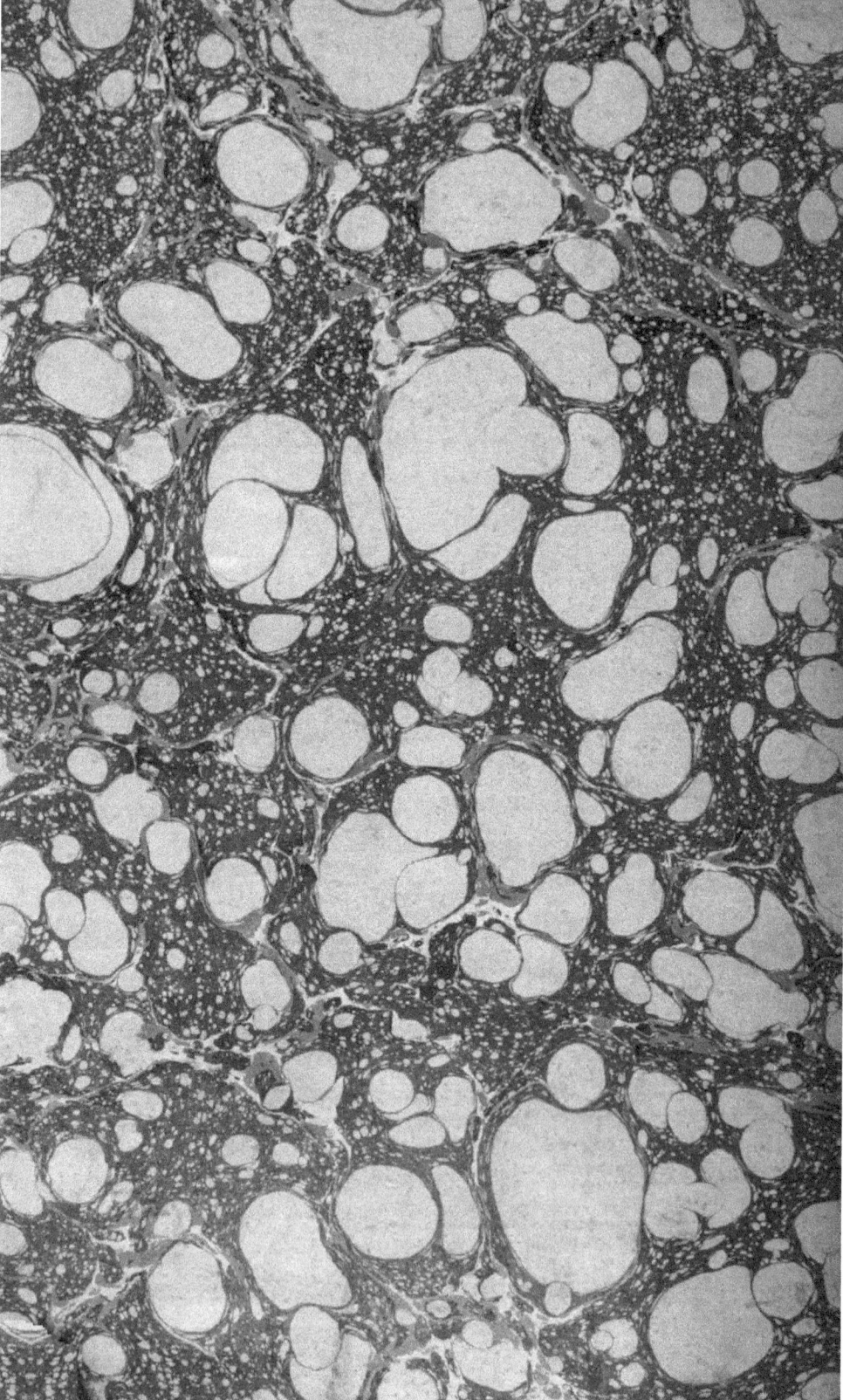

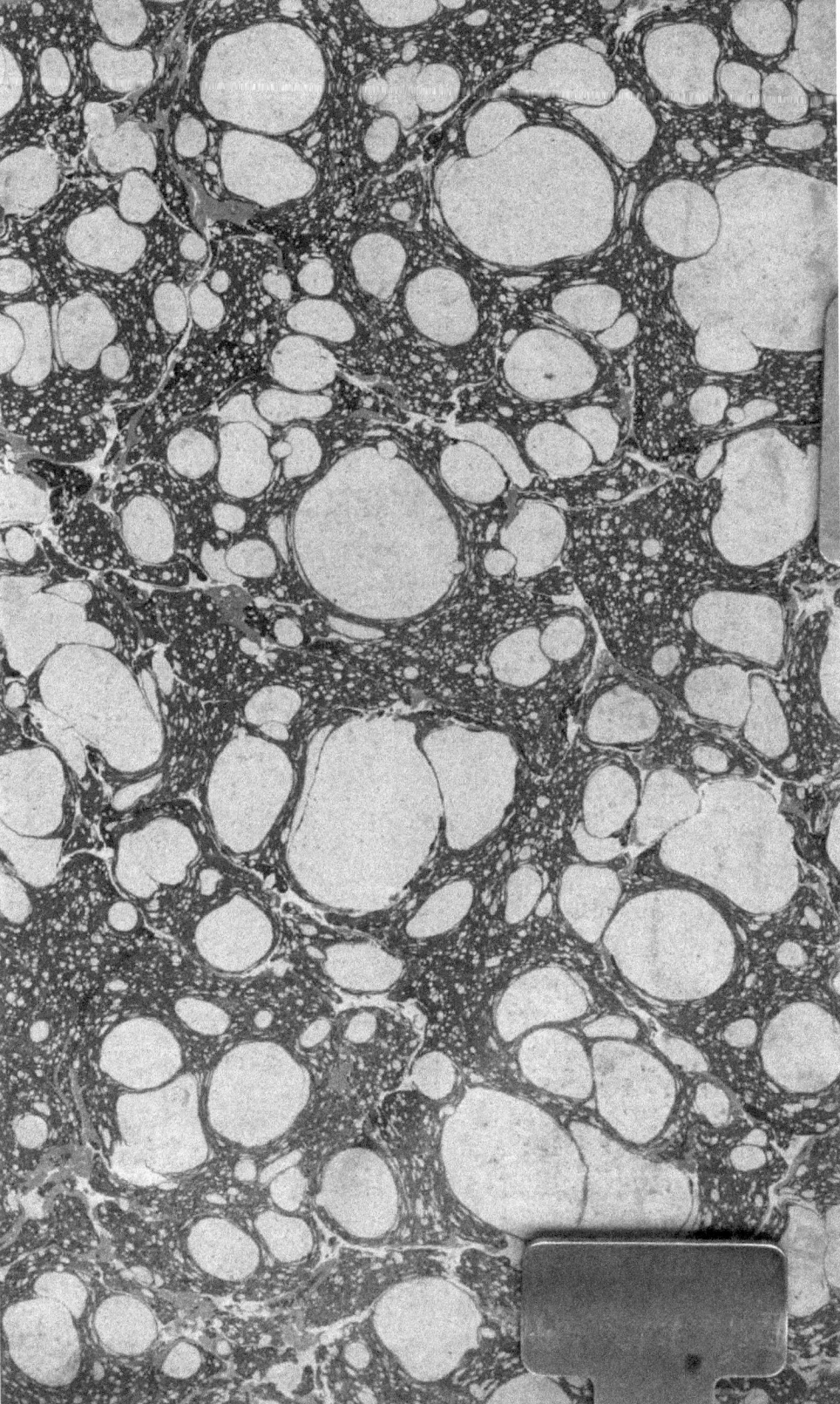

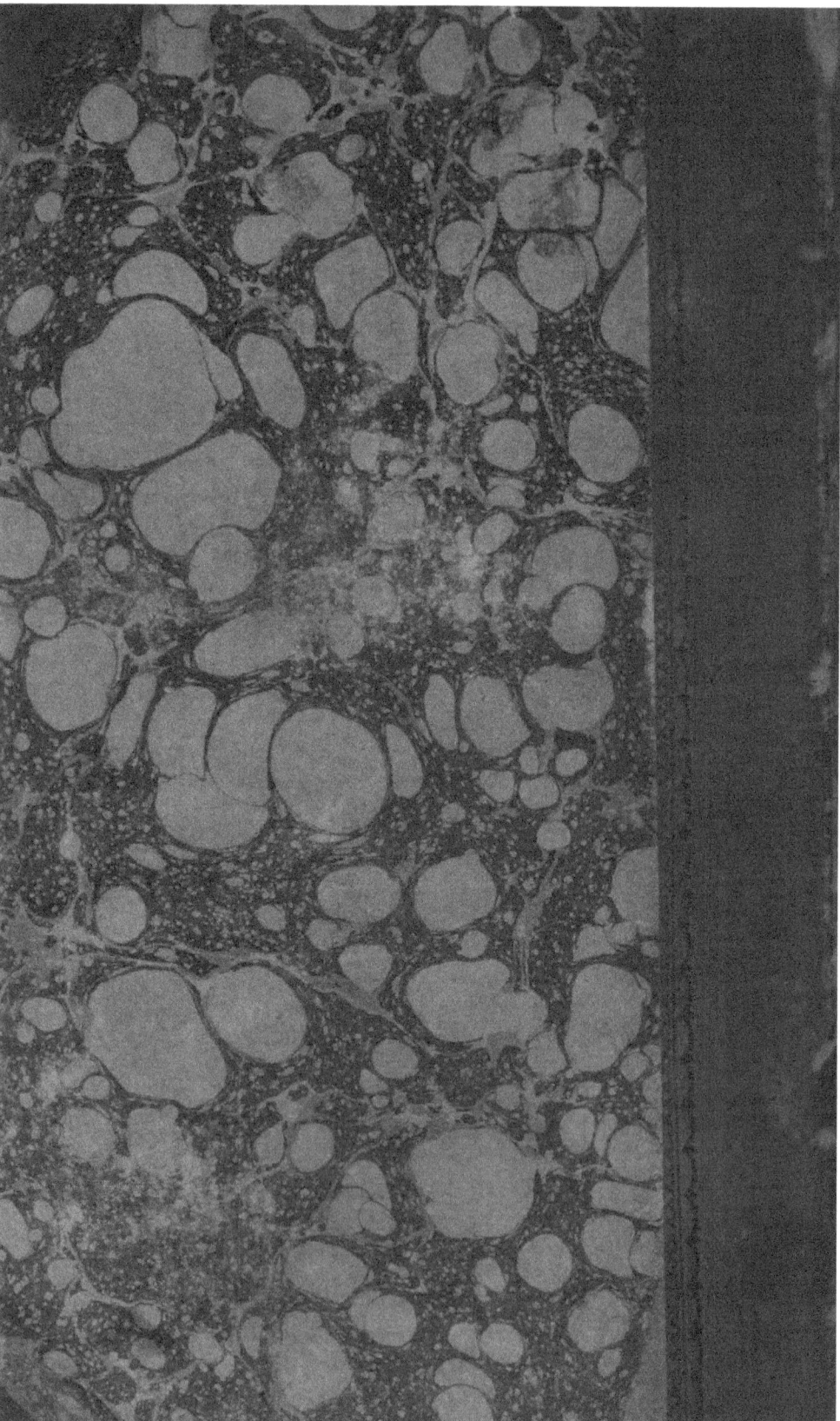

Check Out More Titles From HardPress Classics Series In this collection we are offering thousands of classic and hard to find books. This series spans a vast array of subjects – so you are bound to find something of interest to enjoy reading and learning about.

Subjects:
Architecture
Art
Biography & Autobiography
Body, Mind &Spirit
Children & Young Adult
Dramas
Education
Fiction
History
Language Arts & Disciplines
Law
Literary Collections
Music
Poetry
Psychology
Science
…and many more.

Visit us at www.hardpress.net